CONTENTS

KT-436-629

genetics

Morton Jenkins

For UK orders: please contact Bookpoint Ltd, 130 Milton Park, Abingdon, Oxon OX14 4SB. Telephone: (44) 01235 827720. Fax: (44) 01235 400454. Lines are open from 09.00–18.00, Monday to Saturday, with a 24 hour message answering service. Email address: *orders@bookpoint.co.uk*

For U.S.A. order enquiries: please contact McGraw-Hill Customer Services, P.O. Box 545, Blacklick, OH 43004-0545, U.S.A. Telephone: 1-800-722-4726. Fax: 1-614-755-5645.

For Canada order enquiries: please contact McGraw-Hill Ryerson Ltd., 300 Water St, Whitby, Ontario L1N 9B6, Canada. Telephone: 905 430 5000. Fax: 905 430 5020.

Long renowned as the authoritative source for self-guided learning – with more than 30 million copies sold worldwide – the *Teach Yourself* series includes over 300 titles in the fields of languages, crafts, hobbies, business and education.

British Library Cataloguing in Publication Data
A catalogue record for this title is available from The British Library.

Library of Congress Catalog Card Number: On File

First published in UK 1998 by Hodder Headline Plc, 338 Euston Road, London NW1 3BH.

First published in US 1998 by Contemporary Books, A Division of The McGraw-Hill Companies, 1 Prudential Plaza, 130 East Randolph Street, Chicago, Illinois 60601 U.S.A.

The 'Teach Yourself' name and logo are registered trade marks of Hodder & Stoughton Ltd.

Cover photo © Science Photo Library

Illustrations by Techtype

Typeset by Transet Limited, Coventry, England.
Printed in Great Britain for Hodder & Stoughton Educational, a division of Hodder Headline Plc, 338 Euston Road, London NW1 3BH by Cox & Wyman Ltd, Reading, Berkshire.

Impression number   10 9 8 7 6 5
Year   2002

INTRODUCTION

'There but for a series of mutations go I.'

This is a sentiment shared by biologists when they wonder at the variety of living things on our planet. This variety has sparked the imagination of many students and teachers of biology, providing a fascination which lasts for the rest of their lives.

Without mutation there can be no permanent change in a species – there can be no variety, no adaptation and no evolution. To understand how species evolve, we must look at the fundamental causes of mutation and see how these changes that result from mutations are passed on through generations.

The science of genetics is of interest to everyone. We can live long and interesting lives without ever needing to know about the intricacies of most branches of science, but we can hardly fail to reflect sometimes on our own origins. Moreover, in recent years there has been widespread speculation about changes that could affect the genetic material inherited by our descendants. For example, rates of mutation may increase if the amount of ionising radiation in the environment rises. Such issues are difficult to evaluate objectively without a sound knowledge of the basics of genetics. Terms such as 'genetic engineering' and 'cloning' have become commonplace but are not always used accurately, leading to some remarkable distortions of truth and misconceptions with a science fiction flavour in the media. Genetic engineering developments often involve ethical issues, and this is why balanced, accurate accounts of such developments are more difficult to come by. Perhaps we are in the middle of a genetic revolution comparable to the Industrial Revolution of the eighteenth and nineteenth centuries.

The science of genetics is essentially a product of the period since the 1880s, but the rate of scientific advance in applied genetics during the last decade of the twentieth century has been truly remarkable. This book is intended to show you the basic principles of genetics so that you can not only understand the concepts discussed within the following pages but also interpret media accounts of new developments in genetics as they appear. You will then be able to consider the science and ethical considerations behind new techniques more objectively.

It is the aim of this book to introduce the principles of genetics in a way that you will be able to understand whether or not you have previous knowledge of the subject. Owing to the rapid advances of this science in recent years, it has become impossible to cover all new developments in detail. This account is intended to serve only as an outline and may be used as a starting point for studies in greater depth. The use of technical terms has been reduced to the minimum, and such terms are explained either at the point where they are first used or in the glossary. For a true understanding of genetics it is essential to grasp the fundamental concepts of the cells or building blocks of all living things, because these contain the material that passes information from one generation to the next. This book begins with an account of how genetic material behaves in cells, leading on to what the genetic material is and how it carries the information that controls the development and evolution of our species.

The way in which characteristics are inherited seems to be identical in all organisms. From viruses to humans, there is a universal code in the genes that carries a message from one generation to the next. When you look at the living world, remember that the only criterion for the success of a species is its ability to pass on its genes to a new generation. Those species that fail to meet this criterion will survive at best only as fossils. Survival of life itself depends on the genes that connect one generation the next. The study of this mechanism for the continuation and evolution of life is the science of genetics.

1 | THE BASICS OF GENETICS

Cells – the units of life

The discovery of cells by Robert Hooke

The impact of the microscope and Robert Hooke's publication *Micrographia* on the scientific world of the seventeenth century is well known. Hooke's observations of cells in cork described in this book in 1665 were later followed by the discovery that other biological materials were also made up of cells.

In the middle of the seventeenth century, Robert Hooke was appointed Curator of Instruments to the Royal Society of London, a scientific organisation with a reputation based on high achievement. Hooke had not received a normal schooling due to ill health in his early years. At Oxford, Hooke's precocity and intelligence attracted the attention of the famous scientist Robert Boyle, who sponsored his appointment to the Royal Society.

The curator's job was to provide the audience at the Society's weekly meetings with a variety of entertaining scientific demonstrations. Hooke's problem was that the Society was composed of the elite scientists of the time, most of whom lacked any sense of humour and were almost impossible to impress. Hooke had the bright idea of demonstrating his latest novelty – one of the first microscopes, which he had built. This was much better than any of the crude models available at the time. In the mid-1600s the scientific world was very concerned with lenses, which at that time were generally no more than rounded pieces of polished clear glass. Lenses were used by book restorers and also by astronomers to observe the moon and stars more clearly. Hooke decided to let the Royal Society's members use his instrument to observe almost any small objects that he could lay his hands on. The scientists were most impressed and were quite content to gaze through his remarkable construction at each meeting.

Week after week they examined minerals, textiles, small plants and insects. During one meeting Hooke cut a very thin slice of cork and placed it under his lens. Those present were interested in the properties of cork because they were curious to understand how a solid substance could float. He focused his lens, but saw nothing because the reflected light used by the microscope was not strong enough. However, once he had devised a way of passing light through the cork he made a discovery that founded a whole new science – the science of cytology, or the study of cells. What he saw reminded him of a honeycomb – an interlocking arrangement of tiny boxes or cells.

In 1665 Robert Hooke published his famous book entitled *Micrographia*. It was one of the most important scientific books of all time, recording his observations in minute detail. He could make no photographic records, so his written and diagrammatic descriptions had to be meticulous. An account from his book is given below. It is of course written in seventeenth century English, but remains a masterpiece of descriptive observation.

Observation XVIII. Of the Schematisme or Texture of Cork and of the Cells and Pores of some other such frothy Bodies.

I took a good clear piece of cork and with a penknife sharpened as keen as a razor, I cut a piece of it off, and thereby left the surface of it exceedingly smooth, then examining it very diligently with a microscope, methought I could perceive it to appear a little porous; but I could not so plainly distinguish them as to be sure that they were pores ... . I with the same sharp penknife cut off from the former smooth surface an exceedingly thin piece of it, and placing it on a black object plate ... and casting the light on it with a deep plano-convex glass, I could exceedingly plainly perceive it to be all perforated and porous, much like a honeycomb, but that the pores of it were not regular ... these pores, or cells, were not very deep, but consisted of a great many little boxes, separated out of one continued long pore by certain diaphragms ... . Nore is this kind of texture peculiar to cork only; for upon examination with my microscope, I have found that the pith of an elder, or almost any other tree, the inner pulp or pith of the cany hollow stalks of several other vegetables: as of fennel, carrots, daucus, burdocks, teasels, fern, some kinds of reeds, etc., have much such kind of schematisme as I have lately shown that of cork.

Although Hooke was the first person to describe cells, he did not realise that the most important part of the cells was missing from his descriptions. What he was observing was just the outer boundary or **cell wall**. The empty shells he saw had once held active living materials. There was no suggestion in Hooke's report that his discovery would be of great biological importance – he did not realise that he had seen the building bricks of life itself.

The Royal Society published Hooke's *Micrographia*, and because King Charles II was the patron of the Society Hooke began his book with a dedication: 'To the King, Sir, I do here most humbly lay this small present at Your Majesty's Royal feet.' It may have seemed a small present for a king but became a priceless gift to the human race.

The discovery of the nucleus

Although animal cells have no cell walls like the ones Hooke described, both plant and animal cells are alike in containing living material, the **cytoplasm**. In the cytoplasm is the nucleus, which is the controlling centre of the cell, and as we shall see, the keeper of the secrets of genetics.

The nucleus was first described in 1700 by Antoni van Leeuwenhoek, a Dutch draper who spent his spare time grinding lenses and who made the best microscopes of his day. He placed some fish blood on a clean piece of glass and through his microscope saw little oval particles – red blood cells. He noticed that in some of these particles there was 'a clear sort of light in the middle'. Without knowing it, he had observed the nucleus of a living cell. Ironically, if he had examined the red cells of human blood or the blood of any other mammal, he would not have seen a nucleus because mammalian red blood cells are among the very few cells that do not have nuclei.

In 1781 Felix Fontana found oval bodies inside the skin cells of an eel. Once again the nucleus was described, but its importance was not recognised. It was the Scottish botanist Robert Brown in 1883 who first established the idea that a nucleus is a normal part of a living cell, by examining hundreds of cells of various plant species. Although better known to chemistry students for his description of the Brownian motion of particles, Brown was the first to recognise the cellular nucleus as the central part of a cell. Observations on animal cells were more difficult than on plant cells because of their greater variety of shapes and sizes, but many microscopists of the early nineteenth century subscribed to the idea that animal cells too have nuclei.

The cell theory

In 1839 two German scientists, Theodor Schwann and Matthias Jakob Schleiden, confirmed earlier observations that the cell is the basic unit of life in both plants and animals. This is called the **cell theory**. A number of biologists before Schwann and Schleiden had described the cellular organisation of plants and animals, but the researches of these two men crystallised the basic concept and started a revolution in cellular studies.

The cell theory is probably the most important biological generalisation of the first half of the nineteenth century. It has grown in importance and is central in the continued development of modern biology. After publication of the cell theory, Schwann became professor at the University of Louvain and, after nine years, moved to the University of Liege. He was known as an outstanding experimenter and an excellent teacher. Schleiden was a successful lawyer whose interest in science was so compelling that he gave up his law practice and, after graduating in medicine, devoted himself mainly to plant science. It was Schleiden who encouraged Schwann to finally develop and publish the cell theory in a research paper.

The two collaborators made a strange pair. Schwann was gentle and kind, always avoiding controversy, whereas Schleiden was assured, disputative and certain to provoke discussion. They complemented each other in this single joint effort and provided an exemplary model of co-operative research.

Schleiden published first in 1838 and reached two major conclusions. He stated that plants were built up of cells, and that the embryo of a plant arose from a single cell. Schwann carried out the more comprehensive work and first used the term cell theory. He published this theory in 1847 under the title *Microscopical Researches into the Accordance in the Structure and Growth of Animals and Plants*.

The origin of new cells – cell division

About twenty years later Rudolph Virchow, a famous German physician, published a research paper in which he suggested that cells only come from pre-existing cells. He wrote, succinctly, *omnis cellula e cellula* – all cells from cells.

Where a cell exists there must have been a pre-existing cell, just as the animal arises only from an animal and a plant only from a plant. The principle is thus established, even though the strict proof has not yet been produced for every detail, that through the whole series of living forms, whether entire animal or plant organisms or their component parts, there are rules of eternal law and continuous development, that is, of continuous reproduction.

We take it for granted today that all cells result from cell division. The process of cell division can be demonstrated to classes of students in school laboratories with easily obtained materials and with microscopes normally found in schools. In the nineteenth century, however, things were quite different. It took some brilliant studies by the self-taught amateur Wilhelm Friedrich Hofmeister to show the world how cells divide. He worked in his father's business in Leipzig as a publisher and bookseller. In his spare time, rather like Leeuwenhoek, he examined living things using a microscope. There are stories of his obsession with this interest, relating how he would rise at four in the morning to devote time to his microscope before going to work. He had many research papers published in the 1840s and 1850s and, on the strength of his achievements, became Professor of Botany at Heidelberg University in 1863.

Hofmeister carefully noted that before a cell divides into two new cells its nucleus first divides to produce two daughter nuclei. He also recorded that before dividing the nucleus first undergoes changes that result in the formation of small rod-shaped bodies which showed up as coloured structures after the addition of certain stains. He called these rod-shaped bodies **chromosomes** from the Greek 'chromos' meaning colour and 'soma' meaning body.

Many scientists continued this line of research and found that the nucleus of any plant or animal cell changes to form rod-shaped chromosomes when the cell is about to divide. Also, the number of chromosomes is constant according to the species of plant or animal. A misconception that students sometimes have is that the more complex the organism, the more chromosomes it has. This is not true, as you can see from Table 1.1.

Table 1.1 The number of chromosomes in the cells of a variety of species

Animals		Plants	
Species	**Chromosome number**	**Species**	**Chromosome number**
Amoeba	50	Garden pea	14
Starfish	36	Carrot	18
Fruit fly	8	Wild rose	14
Housefly	12	Tobacco	48
Goldfish	94	Lettuce	18
Alligator	32	Magnolia	20
Chicken	78	Marijuana	20
Dog	78	Onion	16
Chimpanzee	48	Redwood	22
Human	46	White ash	46

The process of cell division – mitosis

It took 200 years from Hooke's first observations of cell walls to knowledge of the details of nuclei and cell division. The science of cytology may have started slowly, but perhaps the 1880s could be described as the end of its beginning, probably because the microscope was reaching a high degree of precision. By 1880 there was general agreement that cells in both plants and animals are formed by equal division, with the nucleus always dividing before the cell. Initially there was disagreement about the behaviour of the chromosomes. Some scientists thought that each chromosome divided in half across the middle. However, another brilliant German contribution to cytology came in 1882 when Walther Flemming published his discoveries. His work was outstanding for a variety of reasons. First, he had access to the most sophisticated microscopes of his day as German technology in the field of precision optical engineering was second to none. Second, new dyes were being developed for staining biological material for study under the microscope and the combination of these two advances enabled Flemming to carry out work which would have been impossible only a generation before.

Flemming examined a great variety of animal and plant cells and selected those that showed the details of cell division most clearly. It is much easier to see detail in dead cells than in living cells because dead cells can be treated with a variety of dyes which show up various structures clearly. The problem is that you cannot be sure that what you are looking at in a stained dead cell is the same in a living cell – perhaps it never existed in the living cell and was produced by the artificial staining process. You can use the analogy of a hen's egg. When it is boiled, many changes take place like those that happen when cells are prepared and stained. You could study a boiled egg and reach some conclusions regarding its structure, but to what extent would these conclusions be true of a living egg?

Flemming never relied on what he saw in dead cells unless he could observe the same in living cells. He also realised that some of the structures that show up clearly in stained dead cells might not be natural – they could be produced by chemical reactions between the dyes and the cellular structures. As a result his meticulous and careful observations have stood the test of time and many biologists have since confirmed them.

He observed that in the process of dividing, a cell's nucleus passes through an orderly series of changes, which he called **mitosis**. The events were approximately the same in all the animal and plant cells he studied. Two daughter cells are produced from a parent cell, and both daughter cells are identical with each other and with the parent cell in all respects. This means that their chromosomes are identical in number and in structure. Mitosis is the method of cell division by which new cells are produced during growth.

Flemming distinguished various phases in the process of mitosis, shown in Figure 1.1. In animal cells, a small structure near the nucleus called the **centriole** divides (A) and separates (B), giving out a radiating system of protein fibres. The chromosomes take on their characteristic rod-shaped appearance. The centrioles travel to opposite poles of the cell and then become connected by fibres forming a spindle-shaped arrangement (C). The chromosomes arrange themselves along the equator of the spindle (D), midway between the centrioles. At this stage each chromosome is made up of twin threads joined only at one place, the **centromere**. After arranging themselves at the equator, the twin threads of the chromosomes separate (E). Now one complete set of daughter chromosomes migrates to each pole, and in co-ordination with this the entire cell divides, animal cells by pinching in two (F), plant cells by building a new cell wall. The

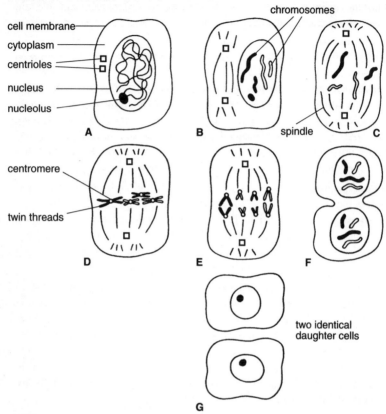

**Figure 1.1 The stages of mitosis, in which a cell divides
to form two identical daughter cells, each with the same
chromosomes as the parent**

daughter chromosomes disappear from view as a new nucleus forms in
each daughter cell (G). The whole process usually takes about three hours,
though it can happen more quickly or more slowly than this.

Sexual reproduction and sex cells

The result of the remarkable process of mitosis is the production of two
daughter cells which are duplicates of the parent cell. An even more
remarkable process occurs when whole organisms rather than single cells

reproduce. Not only must new cells be made but they must be successfully grouped together to form the structures of the new organism.

The starting point for a new organism is the fusion of a single male sex cell with a female sex cell. This fusion is called **fertilisation**. The human male sex cell is the **sperm** which is too small to be seen without a microscope. The sperm, first described by Antoni van Leeuwenhoek, as a tadpole-shaped cell so small that millions could fit on the head of a pin! In comparison the female sex cell or **egg** is gigantic, and just visible to the naked eye. The egg was first described by another Dutchman Regnier de Graaf when he observed ovaries from small mammals using a microscope.

To suggest that two such minute structures as the sperm and egg can contain all the genetic information needed to grow a new human being appears incredible. It is not surprising that early scientists in this field invented other explanations. It is easy for us to criticise these pioneers today, but remember that in the eighteenth century mystics far outnumbered logical thinkers. Ignorant of fertilisation and of the fact that both the egg and sperm are needed for procreation, they conjured up explanations in which a sex cell contained a mini version of the parent. The Italian Marcello Malpighi (1628–94) suggested that the egg carried a miniature model of the parent. He had examined the stages of development of a chick embryo in a fertilised hen's egg, and his logic was also based on the fact that a flower bud contains a preformed flower. Others including the Dutchman Jan Swammerdam (1637–80) let imagination get the better of them while using their microscopes and thought that they saw tiny embryos in human sperm.

Toads in underpants

In 1780 a fascinating experiment was devised by an Italian abbot, Lazzaro Spallanzani (1729–99). He was convinced that the egg carried an already formed embryo. To resolve the egg versus sperm controversy, he took frogs and toads from the river valley of the Po in his native Tuscany at mating time, and fitted the males with little pairs of silk pants! As a result, he found that the eggs produced by the females remained sterile. He then filtered the seminal fluid shed by the males through filter paper, removing the sperm from the fluid, and mixed the filtrate with the eggs. Nothing happened. By these experiments Spallanzani had in fact demonstrated that sperm are needed to fertilise eggs. However, he did not accept his own results. He had the preconceived idea that only the egg was needed and, in ignoring his observations, demonstrated how not to draw logical conclusions from a scientific investigation.

The basic idea of fusion between egg and sperm was not accepted until the mid-nineteenth century when the process was actually observed. In 1875 a German, Oscar Hertwig, was the first to prove that the nuclei of the egg and sperm fused, and that only one sperm could fertilise an egg.

Biologists today take the fact of fertilisation for granted – indeed, a whole new science of embryology has grown from these revolutionary beginnings. The early observations demonstrated that a new life begins at the moment of fusion between the nucleus of the egg and the nucleus of the sperm. After this the organism grows by continued cell division.

The chromosomes in the gametes

The number of chromosomes in the nucleus is normally constant for any species (see Table 1.1, page 8). However, the sex cells (the **gametes**) do not have the same number of chromosomes as body cells. If they did, a human egg and sperm each having 46 chromosomes would fuse to form a new cell containing 92 chromosomes. Successive generations would have more and more chromosomes until by the end of the tenth generation the individual would have cells each containing 23 552 chromosomes! This is obviously not the case.

A special kind of cell division takes place in the formation of the gametes which halves the normal number of chromosomes. The normal number is thus restored when the two nuclei of the sex cells fuse. In normal body cells the chromosomes are in pairs. In the sex cells, only one member of each pair is present.

This process of cell division is called **meiosis**. It was first suggested by yet another German biologist, August Weismann (1834–1914). He put forward his idea during a meeting of the British Association for the Advancement of Science in Manchester in 1887. Weismann suffered from poor eyesight, so it was difficult for him to use a microscope. However as a result of logical interpretation of data he was able to explain how chromosome numbers might stay constant from one generation to another. He considered two facts:

1 In successive generations, individuals of the same species have the same number of chromosomes.

2 In successive cell divisions, the number of chromosomes remains constant.

He realised that statements (1) and (2) cannot both always be true. So he suggested the idea that there must be a kind of cell division in which the chromosome number is halved. For example, in humans the normal number of chromosomes is 46. He assumed that during the formation of eggs and sperm the chromosome number is halved so that the egg has 23 chromosomes and the sperm has 23 chromosomes. When egg and sperm combined, the sum would be 46 chromosomes again. The fertilised egg would then divide by mitosis and every cell so formed would have 46 chromosomes.

Weismann's prediction was soon verified. Biologists carefully studied the cells that developed into sex cells and found that just before the mature sex cells are formed the chromosome number is indeed halved. This happens in all organisms that reproduce sexually.

Five years later Weismann made another very important suggestion. He wanted to account for the fact that different offspring from the same parents are not identical. He proposed that inherited characteristics are reshuffled in the formation of the sex cells. This would mean that different sex cells in one organism would contain different combinations of hereditary characteristics. He added that this reshuffling must occur during the cell division that halves the chromosome number. Thus each sperm and each egg is unique, and each fertilised egg cell is also unique. The advantage of sexual reproduction lies in this uniqueness of the fertilised egg, making each new life just a little different from its parents or siblings. Such variation among offspring is likely to produce one or more individuals that can adapt to changing conditions in the environment and thus survive.

Eventually the new organism will reach maturity and will itself make a contribution to the survival of its species by reproducing. The cycle of life will have made a full turn.

Meiosis

The process by which sex cells are formed is a type of cell division called **meiosis**, shown in Figure 1.2. During meiosis elaborate chromosome manoeuvres take place. In the early stages the chromosome pairs move towards each other (B) and appear united along their whole lengths (C). At this point the chromosomes are so close together that the nucleus looks as though it has only half its normal set of chromosomes.

The chromosomes now shorten and thicken and divide in two lengthwise to form twin threads. The double pairs now begin to separate (D), though the separation is not complete because at one or more points along their length the threads of the double chromosomes remain in contact. Odd shapes are produced as the chromosomes cross over each other. These points of contact or crossover points are called **chiasmata** (from the Greek 'chi' or cross). Next the double chromosomes separate from each other (E) and move to opposite poles of the cell (F). A nucleus then forms at each pole, and the cell divides into two (G). At this point there are two daughter cells each containing chromosomes made up of twin threads. Finally these double chromosomes separate and each daughter cell divides into two, as in mitosis (H–K). This results in four cells in all, each with half the normal number of chromosomes. These cells are the sex cells, the sperm or eggs.

The origins of genetics

Life begins with the joining together of two tiny blobs of matter – the father's sperm and mother's egg. These two sex cells contain all the **genes** or instructions needed to control the physical development of the individual. This fact is one of the marvels of life on Earth. The understanding of how genes carry information and how they control the development of cells is a story of staggering and wonderful complexity.

The first biologists to study inheritance described the outward features that are passed from one generation to the next. They formulated various complex rules which enabled them to explain, for example, why more men are red–green colour-blind than women. However, although these pioneers were able to predict a trend or pattern of inheritance, they could not understand *how* features are inherited. In order to explain how a biological event happens we must eventually understand the working of

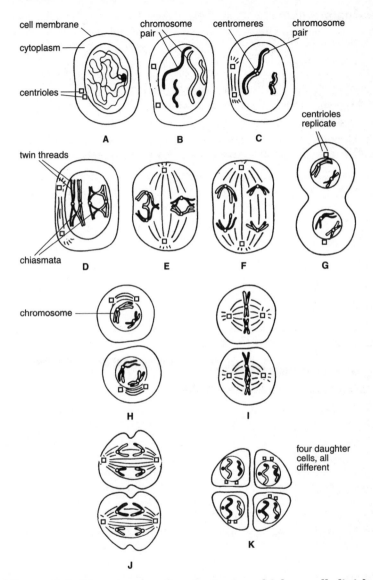

Figure 1.2 The stages of meiosis, in which a cell divides to form four daughter cells with half the number of chromosomes of the parent cell

the molecules that make up all living things. Traditional biologists have observed the structure and functioning of whole living things. Biochemists have studied how the molecules of living things react. Today molecular biologists combine the two approaches and attempt to describe the functioning of living things in terms of the molecules that make them up. Eventually they may be able to explain all living processes in this way, and even assemble living matter. The development of this field is explained here in relation to genetics.

In the beginning – breeding experiments

The word **genetics** was coined by William Bateson in 1907. Genetics is the study of how parents pass on characteristics to their offspring. For hundreds of years farmers knew that if they bred from cows that gave most milk, or from the wheat with the largest grains, they were likely to get these useful features again.

Did you know?

Did you know that there was once an experiment to recreate Leonardo da Vinci. Leonardo was the illegitimate son of Piero and a peasant girl named Caterina. Piero was a notary (an educated person authorised to draw up deeds and contracts) who lived in Vinci, Italy. The third wife of Piero later had a son who they named Bartommeo. He idolised his half-brother Leonardo despite being forty-five years younger. After the death of Leonardo, Bartommeo attempted an amazing experiment. He studied every detail of his father's relationship with Caterina. Then Bartommeo, who was also a notary, returned to Vinci and found another peasant girl who seemed similar to Caterina. He married her and they had a son whom they called Piero. Strangely the child actually looked like Leonardo and was encouraged to follow in the great man's footsteps. The boy was an accomplished artist and was on the way to becoming a talented sculptor when he died, thus ending the experiment!

The results of breeding experiments did not always turn out as expected, however. Sometimes the animals selectively bred on farms had useful features; at other times they did not. Desirable features which had not appeared in the first generation of offspring as expected sometimes reappeared in the second generation. Why?

The answers to questions about inheritance plagued many scientists. As we shall see, the mystery was finally solved by an Augustinian abbot Gregor Johann Mendel (1822–84) in a monastery in Brunn, Moravia, now the Czech Republic. However, as early as the seventeenth century there was recorded evidence to suggest that people were aware of genetic puzzles. A manuscript dated 1669 by Johann Joachim Becher states:

When a black cock pigeon and a white hen pigeon unite, the young birds of the first generation are usually some of them entirely black and others entirely white; and it is only when we allow some of these blacks and some of these whites to unite that we get young birds which are spotted black and white. In arboriculture nature achieves similar results, for when there occurs a union between trees bearing white and red fruits respectively, spotted fruits only appear after the second crossing.

Around the time of Mendel's birth in 1822, John Goss of Devonshire had recorded results of experiments with garden peas in the *Journal of the London Horticultural Society*. He had crossed plants that produced green seeds with plants that produced yellow seeds. The cross produced only plants that produced yellow seeds in the first generation. When he cross-bred the first generation Goss was astonished to see three kinds of plants – one with only green seeds, one with only yellow and the third with both green and yellow seeds in the same pod.

Another notable observer of genetics was Charles Darwin, the founder of the theory of evolution. He also crossed varieties of garden peas, but in addition recorded results of crossing experiments with antirrhinums (snapdragons). Like others before him, Darwin failed to work out a logical explanation for his genetic experiments.

Gregor Mendel's contribution to genetics

The abbot's garden of inheritance

The abbot Gregor Mendel was a particularly brilliant man. He is often portrayed as a genial old monk who accidently discovered important laws governing the patterns of inheritance while pottering around his monastery garden. In reality, Mendel was probably the first mathematical biologist.

Gregor Mendel was christened Johann Mendel, of peasant ancestry in Moravia. His family was poor and struggled to provide his early education. Mendel first attended the village school and then studied at the Gymnasium in Troppau. Despite several illnesses throughout his schooling he was top in almost all his classes and in 1841 progressed to the College in Olmuetz to study philosophy. He graduated in 1842 and then, after being influenced by his uncle, Mendel took up holy orders. At the age of twenty-one he was ordained, taking the name of Gregor by which he has been known ever since.

Early in his life Mendel trained himself in the sciences and became an accomplished observer of natural history. In order to make a living as a young man he worked as a part-time science teacher in a local school. His superiors at the school realised that he had potential as a teacher and suggested that he became qualified. Mendel attempted the qualifying examination but failed to reach the high standard of excellence demanded at that time. Fortunately the order of monks to which he belonged believed in his ability to succeed and in 1851 sent him to the University of Vienna to study natural sciences and mathematics for two years. The science of genetics probably owes an unpayable debt to the person who made this single decision.

On returning to the monastery Mendel decided to begin serious research in breeding experiments. For the next fourteen years he worked patiently in the monastery, making one of the two rooms he lived in into a laboratory. This became a menagerie, containing birds, mice and even a porcupine and a fox! In the garden he kept bees, recorded meteorological data and began his experiments crossing plants.

Mendel applied his logic and investigative skills in earnest and developed new varieties of fruits and vegetables, while keeping abreast of the latest discoveries in plant hybridisation (cross-breeding). The success of his research was probably because he brought three scientific disciplines to his methods – philosophy, mathematics and biology.

In 1865, while the great Charles Darwin was still puzzling over the riddle of how characteristics are passed from one generation to the next, Mendel presented a single research paper to the members of the little known Brunn Natural Science Society. About forty people heard Mendel describe his work, and also came the following month to a second meeting when he completed his account. He was politely applauded and then the assembled group of experts ignored him and had a lively discussion on the topic that was the flavour of the month – Darwin's theory of natural selection. Mendel's paper was published in the following year in the society's journal, the *Transactions of the Brunn Natural History Society*. Its title was 'Versuche uber Pflanzenhybriden' ('Experiments on Plant Hybridisation'). Despite the fact that the journal was circulated to 120 or so other learned societies throughout the scientific world, no one took any notice. The work for which Mendel was later to become so famous was met with a resounding silence.

In 1868 he was made abbot of the monastery and of necessity spent most of the rest of his life in the management of his monastic duties. His lifestyle changed and because of the need to entertain prominent guests he put on weight. He was constantly trying to reduce this and it is said that his slimming methods included rolling on the floor of his bedroom, rising at four in the morning and adopting a liquid diet. None proved successful. He took to heavy smoking, and smoked twenty cigarettes a day by the time he died at the age of 62. In the last years of his life he was unable to proceed with his research and endured long arguments over taxation of properties and other trivial matters compared with his contribution to science. He died from several causes including heart and kidney failure. He insisted in his will on an autopsy because of his fear of being buried alive.

Mendel had lived a full and successful life and was not in any way bitter about the fruitless outcome of his work while he lived. G. von Niessal, a friend of his, wrote: 'Mendel did not expect anything better, but I heard him in the garden express the prophetic words: "Meine Zeit wird schon Kommen" (My time will come)' – a poignant and fitting epitaph.

Even when Mendel died in 1884 his work remained unknown. A reason for this might be that Mendel was many years ahead of his time in his conceptual level of thinking about the nature of inheritance – nothing was known yet about genes or DNA. Also, the scientific world was still busy analysing Darwin's revolutionary *Origin of Species*. However, it is ironic to think that a complete explanation of natural selection depended on Mendel's work!

Like many original thinkers, Mendel died unrecognised. His funeral was attended by those who honoured him as a good priest and as an inspiring teacher, but there could have been none there who realised that they were honouring a great scientist. Today we call him the 'father of heredity' or even the 'father of molecular biology'.

Mendel's laws of inheritance

The statements that Mendel made were as follows:

1 When parents differ in one characteristic, their offspring will be hybrids of that particular characteristic. However, the offspring bear the traits of one parent instead of showing a blend of the traits of both parents. This means that the trait of one parent must somehow be dominant over the trait of the other parent. He called this phenomenon the **principle of dominance**.

2 When a hybrid reproduces, its gametes (egg or sperm) will be of two types. Half will carry the dominant trait provided by one parent and half will carry the recessive trait provided by the other parent. He called this the **principle of segregation**.

3 When parents differ in two or more characteristics, the occurrence of any characteristic in the next generation will be independent of the occurrence of any other characteristic. That is, the tendency towards tallness in plants is not associated with any specific seed colour, for example. The two are inherited independently. Thus any combination of the parental characteristics may appear in the offspring. This is called the **principle of independent assortment**.

Having read these three statements you are possibly as confused as Mendel's original assembly of experts were in 1865! The following account of how Mendel reached his conclusions should clarify things.

Mendel's experiments

Why pea plants?

Mendel carried out a series of meticulously organised investigations and then applied statistical analysis to the results. Using mathematics to analyse observations from biological investigations was an entirely new idea. As on many other occasions throughout the history of scientific discovery, good luck played a part in his studies. In this case the piece of good fortune was Mendel's choice of experimental material – he chose the garden pea *Pisum sativum*. If he had chosen almost any other plant the clear formulation of his laws of genetics would not have been possible, for reasons that we shall see later. What is so special about peas?

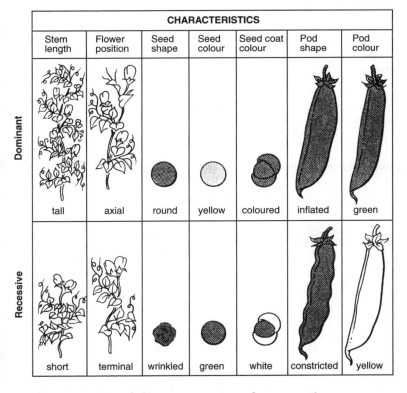

CHARACTERISTICS						
Stem length	Flower position	Seed shape	Seed colour	Seed coat colour	Pod shape	Pod colour
tall	axial	round	yellow	coloured	inflated	green
short	terminal	wrinkled	green	white	constricted	yellow

Dominant (top row) / Recessive (bottom row)

Figure 1.3 Mendel's seven pairs of contrasting characteristics (traits) in the garden pea

Peas were readily available and easy to grow in the large numbers necessary for statistical analysis, and Mendel had developed thirty-four **pure strains** of pea plants. These were families of plants that always produced a certain characteristic – for example, pure-breeding tall pea plants would always produce more tall pea plants when crossed together. The important thing was that these strains differed from each other in very obvious contrasted ways, so it was easy to observe the results of his investigations. Mendel selected seven different pairs of contrasting characteristics, as shown in Figure 1.3.

Pollination in peas

A knowledge of how flowering plants reproduce sexually is essential in order to understand Mendel's work. Our peculiar uses of flowers include decoration for events such as weddings and funerals, and as peace offerings. Plants use flowers for the serious business of sex. The flower is the place where the male sex cell meets the female sex cell. The subsequent changes result in seeds being formed within fruits.

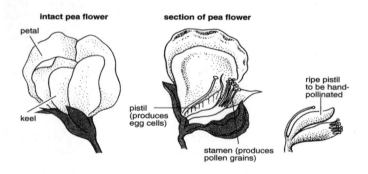

Figure 1.4 The reproductive structures in the pea flower

Each flower of a pea plant contains several stamens, as shown in Figure 1.4, which are structures that produce pollen grains. Each pollen grain contains a male sex nucleus, the plant's equivalent of a sperm. Each pea flower also has a structure called a pistil or carpel, which contains egg cells in its pod-shaped base. These contain the female sex nucleus. The transfer of pollen from the stamens to the pistil is called **pollination** – and this transfer allows the male and female sex cells to meet and fuse for fertilisation to take place.

Pea plants normally self-pollinate and hence self-fertilise. This means that pollen from one flower pollinates the pistil of the same flower. When two strains are crossed, cross-pollination has to be carried out which is not the pea plant's normal way of doing things. Mendel parted the petals of one unopened flower and used a fine brush to take pollen from the ripe stamens, and then dusted the pollen on the ripe pistil inside a second unopened flower. Using unopened flowers meant it was impossible for pollen from an unknown plant to have already pollinated the second flower, and so he could be certain of the genetic make-up of the resulting seeds. He made sure the second flower could not self-pollinate by removing the stamens from it at an early stage in the development of the pea flower.

Mendel's principle of dominance

Mendel's procedure was novel for that period, although today it is the basis of the production of many new varieties of plants. He crossed two true-breeding strains that differed in one single characteristic, for example seed colour.

Mendel gave the symbol P to the pure-breeding parents that he crossed, and the symbol F_1 (**first filial generation**) to the offspring. When the F_1 plants were allowed to self-fertilise, the **second filial generation** produced was called the F_2, and so on through further generations.

When Mendel crossed his original P plants he found that the characteristics of the parents did not blend, as fashionable theory of the time suggested they should. For example, when plants with yellow seeds were crossed with plants with green seeds, the F_1 generation plants were not yellowish green or greenish yellow – they were all as yellow as the original parent. Because of this observation, Mendel called the characteristic that appeared in the F_1 generation the **dominant** characteristic. The green characteristic seemed to have receded into the background, so he called it the **recessive** characteristic. It had been passed along through countless generations of green-seeded plants before this cross, so Mendel reasoned that it could not have just disappeared.

Cross (*P*)		F_1 Generation	F_2 Generation	Actual ratio	Probability ratio
 round	wrinkled	 round	5474 round 1850 wrinkled	2.96 : 1	3 : 1
 yellow	green	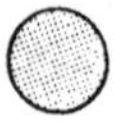 yellow	6022 yellow 2001 green	3.01 : 1	3 : 1
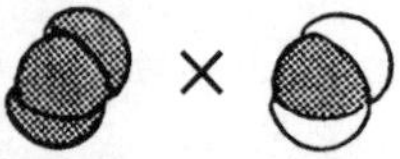 coloured	white	 coloured	705 coloured 224 white	3.15 : 1	3 : 1
 inflated	constricted	 inflated	882 inflated 299 constricted	2.95 : 1	3 : 1
 green	yellow	 green	428 green 152 yellow	2.82 : 1	3 : 1
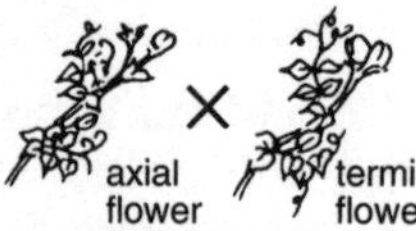 axial flower	terminal flower	 axial	651 axial 207 terminal	3.14 : 1	3 : 1
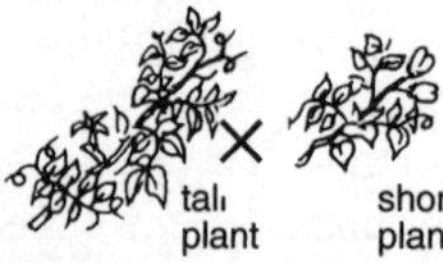 tall plant	short plant	 tall stem	787 long 277 short	2.84 : 1	3 : 1

Figure 1.5 Results of Mendel's monohybrid crosses

Mendel's principle of segregation

In the next stage of his investigations, Mendel allowed the F_1 plants to self-pollinate at random, and found that the missing recessive green characteristic reappeared in some of the F_2 plants! Moreover, the ratio of F_2 plants with the recessive characteristic to those with the dominant characteristic was fairly constant, regardless of which characteristics were involved. The precise results for each of the seven pairs of characteristics Mendel used in his investigations are shown in Figure 1.5.

Too good to be true?

Strangely enough a detailed statistical analysis of Mendel's results made by Ronald Fisher in the 1930s showed that Mendel's figures from his later investigations were, in statistical terms, too good to be true! The implication is that once Mendel had formulated his theory, those results that failed to give the expected ratio were ignored by Mendel or by the gardeners who helped him count his seeds or plants.

Finally, in the third year of his investigations, Mendel allowed the F_2 plants to self-pollinate. He found that all those with recessive characteristics produced only recessive F_3 offspring. Of the F_2 plants that showed a dominant characteristic, one-third produced only dominant offspring, while the other two-thirds produced both dominant and recessive offspring in the ratio of 3 dominant : 1 recessive.

Let us consider one of Mendel's crosses in detail. One pure-breeding strain that always produces round seeds is crossed with a pure-breeding strain that always produces wrinkled seeds. As Figure 1.5 shows, the round strain is dominant to the recessive wrinkled strain. Each plant has two copies of a **gene** that codes for this characteristic. The two alternative forms of the gene are called **alleles**. We use the label **R** for the allele for the round strain and **r** to the allele for the wrinkled strain. The use of a capital letter for the dominant allele and the same lower case letter for the recessive allele is standard practice in genetics. The **genotype** of an organism is the genes it carries. For example, for a homozygous round-pea plant the genotype is **RR**. The **phenotype** of an organism is the external characteristic, such as round peas or wrinkled peas. You cannot tell the genotype from looking at the phenotypes. A round-pea plant may have the genotype **RR** or **Rr**. You can only find out which it is by breeding it with other plants, as we shall see later.

Since both parent plants are pure breeding, we can assume that the two alleles in each plant are identical, that is, in the round parent the genotype is **RR** and in the wrinkled parent it is **rr**. Such plants that have identical alleles for a characteristic are said to be **homozygous** for that particular characteristic. All the gametes produced by a homozygous plant will also carry this same allele.

We can show the cross in a diagram like Figure 1.6. The F_1 generation gains half its alleles from each parent, and since one parent has only **R** alleles and the other only **r** alleles, the F_1 plants are all **Rr**. Unlike the parents, these plants have two different alleles for the characteristic in question and are called **heterozygous** or hybrid. Now when these plants produce gametes, each gamete can carry either an **R** allele or an **r** allele. The combination of alleles present in two gametes that fuse at fertilisation is what determines the characteristics of the resulting F_2 plant. The possibilities are shown in Figure 1.6 in the form of a **Punnet square**.

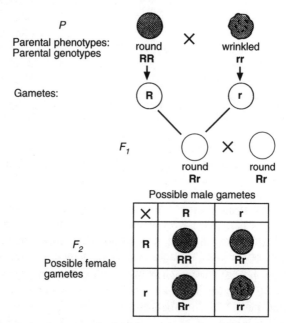

Figure 1.6 A monohybrid cross using a Punnet square

Punnet squares and probability

A Punnet square is a grid named after its inventor R C Punnet in 1905. You can use such squares to predict the results of genetic crosses such as Mendel's. The Punnet square in Figure 1.6 shows the possible allele pairings that can result from a cross between two heterozygous F_1 plants.

The alleles that could be present in the female gametes are on the left of the grid, and the alleles that could be present in the male appear on the top. The alleles from both are combined in the relevant squares of the grid. This shows all the different possible pairings of alleles and hence all the different possible genotypes of the offspring.

Besides showing all the possible allele pairings, a Punnet square also gives the probability for each pairing. That is, it shows how often on average a given pairing is likely to occur. For example, the grid shows a probability ratio of on **RR** to two **Rr** to one **rr**. If there were, say, 100 offspring from this cross, you might expect 25 **RR**, 50 **Rr** and 25 **rr**. However, a small number of offspring is very unlikely to show the exact ratio of offspring predicted by the square. The larger the number of crosses involved, the closer the results will be to the Punnet square probability ratio.

You can see this effect by flipping two coins. In each flip, the coins may land two heads, two tails, or one head and one tail. The probability of one head and one tail is twice as great as that of either two heads or two tails. By chance, a quite different ratio may appear in a small number of flips. However, the more times you flip the coins, the closer the results will be to the probability ratio.

It is important to distinguish between phenotype and genotype in the F_2 probability ratio. What you would expect to see is a ratio of 3 round : 1 wrinkled, as you cannot tell the difference between an **RR** plant and an **Rr** plant by looking at them – they both have round peas. This 3 : 1 ratio in the F_2 generation happens because of the separation of the alleles in the gametes of the F_1 generation and their recombining to form the F_2 generation.

Mendel's investigations involved crossing parents with one pair of contrasted characteristics – we call this a **monohybrid cross**. He showed that whatever it was that actually passed along from one generation to the next could only appear once in a gamete, but twice in a new plant. He called it a germinal unit, but in 1909 the Danish biologist Wilhelm Johannsen named this germinal unit the **gene**. The actual nature of a gene was unknown at this time. This separation of the alleles in the gametes is the basis of Mendel's **law of segregation**: 'Of a pair of contrasted characters, only one can be represented by its germinal unit in a gamete'.

Having established the principles of dominance and segregation, Mendel ruled out the earlier notion that inherited characteristics were always blended in the offspring.

Mendel's principle of independent assortment

In his next series of investigations, Mendel crossed **dihybrid** peas, which were differing in *two* pairs of contrasted characteristics. In one such investigation pure-breeding plants with round (**RR**), yellow (**YY**) seeds were crossed with a strain that had wrinkled (**rr**), green (**yy**) seeds. We saw in Figure 1.4 that round and yellow are both dominant characteristics, so we shouldn't be too surprised to learn that all the F_1 generation had round and yellow seeds. Since one parent was **RRYY** and the other was **rryy**, all the F_1 plants would have to be **RrYy**, heterozygous for both characteristics. When these F_1 plants were allowed to self-pollinate, a random and **independent** assortment of the two characteristics would produce the F_2 generation as shown in Figure 1.7. The results showed a clear 9 : 3 : 3 : 1 ratio.

It was apparent to Mendel from these results that the segregation of germinal units for one characteristic was not affected by the segregation of germinal units for the other characteristic. That is, round peas are not always yellow or always green – the two characteristics round/wrinkled and yellow/green are not connected with each other but are independent. However, it must be pointed out again here that if you were to carry out a single cross to test Mendel's results, you might end up with all wrinkled green seeds in the F_2 generation, just as if you were to flip a coin ten times, it might come up heads each time. You could expect the pattern shown in Figure 1.7 only if you ran sufficient tests for the law of averages to apply.

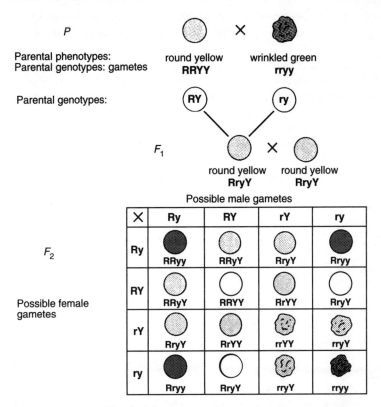

Figure 1.7 A dihybrid cross using a Punnet square

From the results of a very large number of such tests, Mendel formulated his **law of independent assortment**: 'Each of a pair of contrasted characters may combine with each of another pair independently'.

The test cross

If you have a plant that produces round seeds, a dominant characteristic, how can you tell whether it is homozygous **RR** or heterozygous **Rr**? Mendel had carried out numerous tests for determining whether a dominant individual was homozygous or heterozygous, but he soon devised a much simpler procedure, the **test cross**. The dominant plant was simply crossed with a recessive individual. Recessives are always homozygous, so the predictions are straightforward, as shown in Figure 1.8.

1 If the dominant round individual is homozygous, then the test cross becomes **RR** × **rr**, and all the offspring are **Rr** (round).

2 If the dominant round individual in question is heterozygous, then the test cross becomes **Rr** × **rr**, and half the offspring will be likely to be heterozygous round **Rr** and half likely to be wrinkled **rr**.

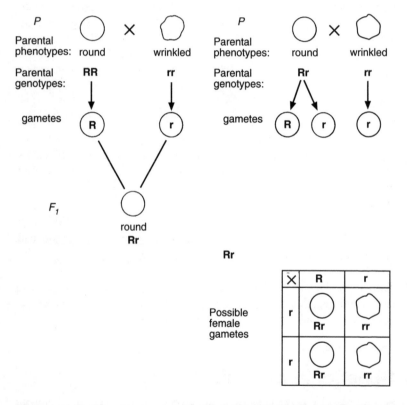

Figure 1.8 The test cross to determine a genotype

The test cross is often applied today to test the pedigrees of plants and animals in agriculture.

The rediscovery of Mendel's laws

Sixteen years after Mendel's death, his work was dramatically rediscovered by three researchers who independently arrived at essentially the same conclusions. The Dutch botanist Hugo de Vries published a report in March 1900 of his considerably lengthy studies of plant hybrids. De Vries was born in Haarlem, the centre of the Dutch bulb-growing industry, and was interested in the origins of the new varieties which frequently added fresh colour to the tulip fields. This stimulated his research into mutations, which we shall pursue in a later chapter. However, his conclusions, he said: 'were, in essentials, formulated long ago by Mendel for a special case .... They fell into oblivion, however, and were misunderstood. According to my own experiments, they are generally valid for true hybrids.' De Vries denied having been aware of Mendel's work until he himself had independently come to the same conclusions.

In April of the same year, the German Carl Correns reported that he like de Vries had independently discovered the principles of dominance and of segregation. Later he realised that Mendel had anticipated him. In fact, his work went further than Mendel and he stated that: 'in the case of a great many pairs of characters, we do not find that either of them is dominant'. We shall return to this issue in another chapter.

Amazingly, in June of the same year, the Austrian botanist Erich Tschermak published the results of his research in which he too had independently arrived at the same conclusions as Mendel. In fairness to the rest, he wrote a postscript to his paper which read: 'The simultaneous discovery of Mendel by Correns, de Vries and myself seems to me particularly gratifying. I, too, as late as the second year of my experiments, believed that I had happened upon something new.'

Genetics in animals

Do animals behave like peas?

Let's look at a monohybrid cross involving guinea pigs. The contrasting characters in this case are black fur and white fur. The allele for black is dominant in these animals.

Consider a cross between a homozygous black guinea pig **BB** and a homozygous white one **bb**. All the F_1 generation are heterozygous black **Bb**. What happens if two of these hybrid animals are crossed? Figure 1.9 shows the expected results.

The probability ratio is one quarter **BB**, one half **Bb** and one quarter **bb** – 3 black : 1 white. These expected results exactly match those in Mendel's crosses of pea plants when he formulated his law of segregation. (Again, remember we are dealing with averages. In any cross between guinea pigs, any combination of black and white offspring *might* result. However, in large numbers of such crosses, the expected ratio would appear.)

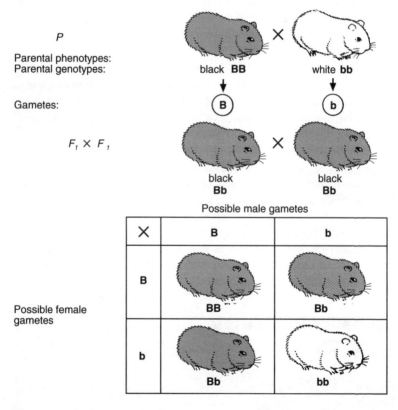

Figure 1.9 A monohybrid cross in guinea pigs

A species jump to humans

Evidently humans cannot be crossed together in experiments like peas. Or perhaps they can! In the eighteenth century Frederick II of Prussia encouraged tall men to mate with tall women. In this way he was able to recruit their tall male offspring to become his guards. However, tallness in men is not the result of just one dominant allele – many genes are involved, together with the effects of the environment such as diet. Also, human genetics depends on people choosing their mates. Even in the most promiscuous, sex is not usually as random a process as it is with peas!

Pedigrees or family trees have been used to trace characteristics through family histories for hundreds of years. Such studies have shown that some human genes do behave like the ones that Mendel studied and have a simple pattern of dominant and recessive inheritance.

Did you know?

Did you know that the word 'pedigree' comes from the French *pied de grue*? It means crane's foot, the shape of which resembles the linked lines of a family tree.

Human genetic disorders

Brachydactyly

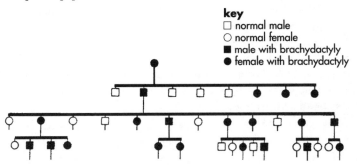

key
□ normal male
○ normal female
■ male with brachydactyly
● female with brachydactyly

Figure 1.10 One of the earliest known pedigrees showing brachydactyly (clubbed fingers)

One of the first recorded pedigrees shown in Figure 1.10 demonstrated the inheritance of short clubbed fingers throughout a Norwegian family. This condition is caused by a dominant allele and is called brachydactyly (from brachy – short and dactyly – digits). If you have normal length digits then you must be homozygous recessive for this character. If your partner also has normal length digits, then any children you might have will be similar. However, if your partner has brachydactyly, then what will be your children's chances of having normal digits? It will depend on whether your partner is homozygous or heterozygous for brachydactyly.

Albinism

Did you know?

Did you know that Noah was probably an albino? 'His hair was white and fine as snow', says the book of Enoch. Maybe this was the first recorded case of a recessive homozygote!

Albinism is caused by a lack of melanin, the normal pigment in the skin, hair and iris of the eye. Because it is caused by a recessive allele, an albino child can be born to normal parents, and if an albino and a normal person had children the chances are that they would all be normal.

Cystic fibrosis

Disorders having a genetic cause are probably more obvious today than they were 100 years ago. In the nineteenth century infectious diseases had a much greater hold, and the effects of genetic disorders were often masked because people died of another cause before their afflictions were diagnosed. Some genetic conditions such as colour-blindness, albinism or dwarfism could be considered mild disabilities or social handicaps, but many genetic disorders have a more disease-like effect and can be very distressing to all concerned.

In the Western world the most common genetic disorder is **cystic fibrosis**. This is caused by a recessive allele, so appears in homozygous individuals. In Britain one person in twenty-five is heterozygous, so is a carrier. Cystic fibrosis is a disabling affliction that formerly always caused death by the early teens. Modern medical techniques can prolong the lives of cystic fibrosis sufferers, but their activities are still restricted. The disorder is characterised by failure to pump salt in and out of cells. Consequently the sweat becomes very salty, but more important, mucus secreted in the chest becomes very sticky and clogs the air tubes. A similar problem arises in the ducts of the liver and pancreas. Pneumonia results from the accumulation of mucus in the lungs, and digestive problems arise because of blockage of the ducts from the liver and pancreas.

Initial attempts to control the disease centred on recognising carriers, who would not show any symptoms as they would be heterozygous. The detection of abnormally high salt concentration in the sweat was one of the first methods used, but proved unreliable because there are many other factors that can lead to this condition. An ingenious test was devised by Barbara Bowman. She noted that blood serum from carriers slowed down the beating of oysters' hair-like cilia, which they use for feeding. The ciliated lining from the windpipes of rabbits also respond to the serum in the same way. These tests were still being used in some laboratories in the early 1980s.

The search for the defective gene on human chromosomes began in 1987, and the gene was finally tracked down by Lap Chee Tsui working at the University of Toronto in 1990. By 1993 screening for the gene became routine – analysis of amniotic fluid identifies fetuses who are afflicted by cystic fibrosis at 18 weeks with a 95% success rate. The next step will probably be gene therapy – the replacement of the defective gene with a normal one (see Chapter 5).

Neurofibromatosis

Another human genetic disorder is **neurofibromatosis** (NF), or Elephant Man disease. Like cystic fibrosis, this is caused by a recessive mutant allele in homozygous individuals, and shows itself in the physical appearance of the afflicted person. It affects about 18 000 people in Britain and its severity varies from just a few coffee-coloured spots on the skin to life-threatening tumours which distort normal features. The symptoms arise from uncontrolled growth of nerve cells.

Huntington's chorea

Most of the human genetic disorders discussed so far have been caused by recessive alleles, so people showing the disorders have to be homozygous for the defective allele. Some disorders are caused by dominant alleles, so heterozygous individuals show symptoms. A particularly distressing

The Elephant Man

The name Elephant Man disease comes from Joseph Merrick, the nineteenth-century Englishman who was called the Elephant Man not so much because of his grotesque appearance but because it was thought that elephants never lay down. Merrick's unfortunate deformed body did not allow him to lie down to sleep, so he slept sitting up. He was thought to have had NF, but X-rays of Merrick's remains in 1996 by Anita Sharma and her team at the Royal London Hospital raised doubts. The radiologists provided evidence that Merrick suffered from a much rarer genetic disease, Proteus syndrome. There have been fewer than 100 cases of this syndrome ever recorded. Merrick died at the London Hospital in 1890, having been rescued from life as a circus freak by doctor Frederick Treves (played by Sir Anthony Hopkins in the film *The Elephant Man*) in 1886. His skeleton was kept in the hospital and is there to this day. Unfortunately Merrick's autopsy records and preserved tissue samples were lost in an air raid in the Second World War. American geneticists isolated the NF gene from human chromosomes in 1990, so geneticists may in future be able to analyse DNA samples from the skeletal remains to see if the NF gene is present.

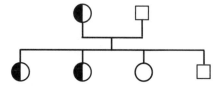

key
☐ normal male
○ normal female
◑ heterozygous female with Huntington's chorea

Figure 1.11 A pedigree showing inheritance of Huntington's chorea

disorder caused by a dominant allele is called Huntington's chorea. The inheritance of this disorder is shown in Figure 1.11. The symptoms include involuntary muscular movement and progressive mental deterioration. The problem is aggravated by the fact that the symptoms do not show until after the age of about 35, by which time individuals could be parents and therefore could have passed on the allele to their children. It is a rare condition with an estimated incidence of one in 18 000 in Britain, and it was not until 1993 that the exact position of the gene was identified on chromosome 4 in humans.

The Fate of Providence?

The history of the study of Huntington's chorea goes back to the seventeenth century. The disorder was introduced to North America in 1630 by two immigrants from Suffolk in England. The name of the disorder is derived from the American George Huntington who first described it in 1872. He came from a medical family and as a boy he used to accompany his father on his professional rounds. On one such occasion, while driving through a wooded lane in Long Island, he saw two women, mother and daughter, both tall and very thin, bowing and grimacing with uncontrolled movements. This incident affected him so much that he vowed to study the problem when he qualified as a medical doctor many years later. Until that time, the affliction was considered to be the dreadful fate of Providence sent to punish sinners.

Human characteristics that obey Mendel's laws

Listed in Table 1.2 are more human traits that are inherited in a simple way as described by Mendel's laws.

Table 1.2

Dominant	Recessive
Dark hair	Blond hair
Non-red hair	Red hair
Curly hair	Straight hair
Abundant body hair	Little body hair
Early baldness in men	Normal
Normal sweat glands	Few sweat glands
Brown eyes	Blue or grey eyes
Hazel or green eyes	Blue or grey eyes
Fee ear lobes	Attached ear lobes
Broad lips	Thin lips
Large eyes	Small eyes
Long eyelashes	Short eyelashes
High, narrow bridge of nose	Low, broad nose
Polydactyly (more than 5 digits)	Normal digits
Syndactyly (webbed fingers)	Normal fingers

2 | A CLOSER LOOK

Building on Mendel's work – genes and chromosomes

Mendel's good luck

When Mendel studied the inheritance of characteristics in plants other than the garden pea, his laws collapsed. His disappointment at this was probably one cause of his depression towards the end of his life. In fact Mendel had been very lucky in choosing first to investigate the particular traits that he did, as each showed obvious contrasting external features and each was either dominant or recessive. This is certainly not always the case with inherited characteristics – many are codominant or incompletely dominant, in other words the contrasted features blend together to form intermediate characteristics in hybrids. Also, genes can be **linked** so that the inheritance of one feature carries with it the inheritance of the other. Some genes are linked to sex, so that inheritance is different in males and females. By pure good fortune, it happened that each of the seven characteristics that Mendel studied in the pea can be inherited independently of the other six. Furthermore, this is the largest number of characteristics in the pea that can be inherited quite independently of each other! Fate was certainly smiling on Mendel during the early days of his research.

Codominance

Exceptions to the principle of segregation

To recap on Mendel's work, two pure-breeding parents with contrasted characteristics were crossed to produce the F_1 generation. The F_1 generation is heterozygous, so has both alleles, but each gamete of the F_1 generation contains the allele for only one of the traits. When these gametes combine to form the F_2 generation, you would expect to see a ratio of 3 : 1 dominant character : recessive character – one homozygous

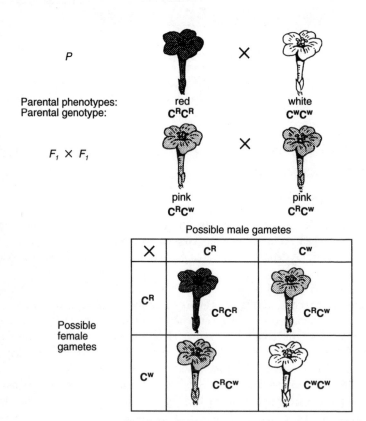

F_2 expected phenotypes: 1 red : 2 pink : 1 white

Figure 2.1 Incomplete dominance in the four-o'clock flower

dominant, two heterozygous and one homozygous recessive, as illustrated in Figure 1.6. This is the basis of the principle of segregation. Inheritance of flower colour in the four-o'clock flower shown in Figure 2.1 appears to destroy Mendel's theory, as instead of 3 red : 1 white or 3 white : 1 red there are 1 red : 2 pink : 1 white in the F_2 generation. However, the genotypes obey Mendel's rule. There is simply an absence of dominance, so that when an allele for white and an allele for red occur together the result is a pink flower. This type of inheritance shows **incomplete dominance**. The

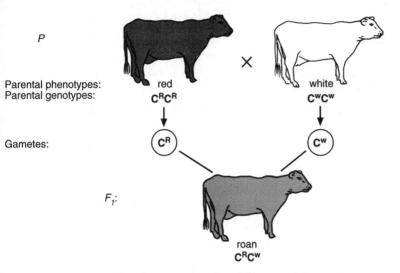

P

Parental phenotypes:
Parental genotypes:

red
C^R C^R

white
C^W C^W

Gametes:

C^R

C^W

F₁:

roan
C^R C^W

Figure 2.2 Codominance in shorthorn cattle

genotype **C^R C^R** means that enough red pigment is made for the full red flower colour to be seen. The genotype **C^W C^W** means that no colour is made at all. The genotype **C^R C^W** means that some colour is made, but only enough to cause the weaker pink rather than red.

A similar pattern called **codominance** is seen in shorthorn cattle. In the cross between a red bull and a white cow as shown in Figure 2.2, the calves are neither all red nor all white. Their coats contain both red and white hairs, a colour called roan. Here the allele **C^R** causes red hairs to be produced and the allele **C^W** causes white hairs. The genotype **C^R C^W** produces a mixture of both types of hairs which show equal dominance.

Did you know?

Did you know that a gene controls animals' internal clocks? In nature, 24-hour clocks are common. Many animals show daily cycles of behaviour, being most active at a particular time. These cycles may be partly due to changes in light and dark periods, but many cycles continue even when animals are kept in constant light or darkness. Golden hamsters maintain cycles of activity of about 24 hours even when they are kept in total darkness.

Scientists at the University of Oregon noticed that one male hamster had an abnormally short cycle when kept in total darkness. Even when it was exposed to 14 hours of light and 10 hours of darkness, the hamster's period of activity began four hours earlier than usual.

The scientists mated the abnormal hamster with females that had 24-hour cycles. After interbreeding the F_1 generation, three groups of hamsters were identified in the F_2. One group had a normal 24-hour cycle, another had a 22-hour cycle and a third group had a 20-hour cycle. On the basis of this evidence the scientists suggested that genes control internal clocks. In this particular case, the 24-hour cycle is controlled by an incompletely dominant allele. The 22-hour cycle takes place in the heterozygote. Can you now explain how the F_2 generation was made up of hamsters with 24-hour cycles, 22-hour cycles and 20-hour cycles in the ratio of 1 : 2 : 1?

Gene linkage

Exceptions to the principle of independent assortment

Mendel's principle of independent assortment says that the inheritance of one pair of contrasted characters is not dependent on the inheritance of another pair of contrasted characters. A cross between double dominant and double recessive parents produces an F_1 generation that shows both dominant characteristics, and in the F_2 generation we expect to see a ratio of 9 : 3 : 3 : 1, as we saw on page 28.

Mendel had accurately described what happened in genetic crosses, but he knew nothing of the nature of genes, or germinal units as he called them. In fact it was not until 20 years after Mendel's work that scientists even became aware that the nucleus of the cell was where the genes are.

Walter S Sutton was a young graduate student at Columbia University, New York, when he published a very important paper in 1903. Chromosomes had been described by Wilhelm Friedrich Hofmeister around 1850. Sutton suggested that chromosomes held smaller genetic particles, each particle accounting for a single trait handed down in reproduction. This was the first time it had been suggested that chromosomes could be carriers of genetic information.

How did Sutton make this breakthrough? He understood Mendel's work thoroughly and he was familiar with the mechanism of meiosis, in which chromosomes segregate to form the gametes. He concluded that chromosomes must carry the materials of heredity. Sutton also thought that the genetic particles must occupy matching positions on each chromosome in a pair. This was later proved to be true.

Soon Sutton recognised a problem. An organism might have a few dozen chromosomes, but might show hundreds or even thousands of inheritable traits. To overcome this problem Sutton suggested that one chromosome must carry a number of genes. Did all these genes segregate independently during gamete formation? Sutton thought not. He believed that they must move in sets on a chromosome. In other words, the genes on a chromosome were linked in a certain order. In 1905 two British researchers, William Bateson and Reginald Crundall Punnet, confirmed Sutton's ideas. They found in experiments on sweet peas that the factors for blue flowers and long pollen grains were linked together – they were not inherited independently of each other.

How could Sutton's theory of gene linkage fit with Mendel's findings? Remember that Mendel's law of independent assortment is based on the fact that each pair of traits segregates and then recombines independently. Surely this could not happen if the genes for two or more of Mendel's seven traits were linked on a chromosome.

Sutton believed that the genes for each of Mendel's traits must be on different chromosome pairs. He could not prove this, but today geneticists know that he was right. The peas Mendel used had seven pairs of chromosomes. By an amazing coincidence, each of the paired traits he studied was determined by genes on a different chromosome pair.

The discovery of sex-linked genes

The black-bellied honey lover

Sutton never proved his theory of gene linkage, but around 1909 another famous geneticist discovered strong evidence to support it. He was Thomas Hunt Morgan, a professor at Columbia University and the Nobel prizewinner for medicine and physiology of 1933. He chose for his studies the fruit fly *Drosophila melanogaster* which translates as the black-bellied honey lover! This fly is about two millimetres long and appears in vast numbers, particularly on over-ripe bananas left lying around in summer.

A professor of entomology, Cylde Melvyn Woodworth, first demonstrated the usefulness of this fly as a laboratory animal at the University of California. More is known about its genetics and its behaviour under laboratory conditions than any other animal, though ironically very little is known about its ecology and behaviour in its natural environment. *Drosophila melanogaster* can complete its life cycle in only twelve days and so can demonstrate hereditary changes quickly and in large numbers in successive generations. Many flies can be kept in a relatively small space and their food is cheap and plentiful. Just as in the case of Mendel's common garden pea, here is another most unlikely species destined for fame.

It all began when Morgan visited Hugo de Vries' experimental garden in Holland at the turn of the century. De Vries had been testing Mendel's laws with evening primroses in 1901. He was surprised to find that, from time to time, a new colour appeared in a pure line of plants. The new colour was inherited in later generations. He called these random changes **mutations**. Morgan was most impressed with what he saw and thought that by studying these flaws in the normal genetic material he would learn more about how genes work. Were there mutations in fruit flies?

In 1909 Morgan began trying to induce mutations in *Drosophila* using a wide range of temperature, various chemicals including radium, and X-rays. We know now that radium and X-rays should have been effective in causing mutations but Morgan's technique for detecting mutations was inadequate, and those mutations that he did notice were not attributed to his treatment. There is more on this on page 74. Success finally came after a year of breeding a culture of *Drosophila* through vast numbers of generations. At last he recognised the kind of mutation that he had hoped for – of all the thousands of normal red-eyed flies there was a male fly with white (albino) eyes.

Morgan bred the single white-eyed male with several of its red-eyed sisters. Imagine the care needed to keep the precious flies alive and to stop them escaping to be lost in oblivion! He obtained 1237 offspring, all with red eyes. This was not surprising because he had assumed that red was dominant to white.

When he crossed the red-eyed F_1 generation together, expecting a ratio of 3 : 1 red-eyed to white-eyed, he was astonished to find 3470 red-eyed flies and 782 white-eyed flies, and every white-eyed fly was male!

In later generations Morgan was able to find white-eyed female flies and so he was able to switch the cross to a white-eyed female with a red-eyed male. All the daughters were red-eyed and all the males were white-eyed. Clearly the gene for white eyes was somehow linked to the mechanism of determining gender. How was gender determined in *Drosophila*? How was the gene for white eyes related to gender?

As long ago as 1881 the French scientist Edouard Balbiani had noted that the salivary glands of *Drosophila* contain giant-sized chromosomes, between 100 and 200 times longer than ordinary chromosomes. However, the significance of this was overlooked at the time by geneticists. In fact, before Morgan no one had studied the chromosomes of *Drosophila* in detail. Morgan and his fellow researchers found that the nuclei of the cells of *Drosophila* had four pairs of chromosomes. Three pairs were the same in both males and females, but in the fourth pair one chromosome was different in the male – instead of being rod-shaped it was hook-shaped. The rod-shaped chromosome in both sexes is called the **X chromosome**, and the hook-shaped chromosome that only occurs in the male is called the **Y chromosome**. These X and Y chromosomes are the **sex chromosomes**. The three chromosome pairs other than the sex chromosomes are the **autosomes**. The fruit fly chromosomes are shown in Figure 2.3.

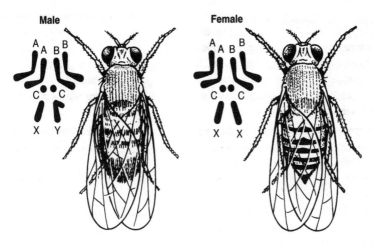

Figure 2.3 Chromosomes in the male and female fruit fly

Figure 2.4 illustrates the inheritance of gender in fruit flies.

Sex chromosomes in other species

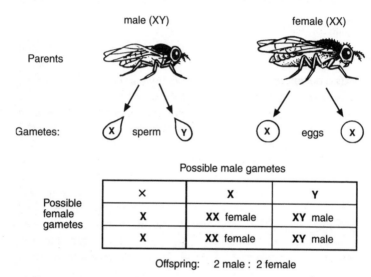

Offspring: 2 male : 2 female

Figure 2.4 The inheritance of gender in fruit flies

These findings explained what determines gender in fruit flies and as we shall see the principle holds true for humans as well. Two X chromosomes (XX) determine female characteristics. An X chromosome paired with a Y chromosome (XY) determines male characteristics. It is the same for humans and most other animals, with the exception of birds and butterflies where the male is XX and the female is XY. The reason for this is still a mystery! In some animals the male has no Y chromosome, so XX produces a female and X results in a male. Where plants have separate sexes, they too have sex chromosomes.

Gender in humans

Each of your body cells have 22 pairs of autosomes and one pair of sex chromosomes. In the 1980s there were several interesting findings regarding gender inheritance. The cells in women have tiny specks in their nuclei called **Barr bodies**. These are named after Murray Barr who first described them in nerve cells in 1949. Also, white blood cells in women show a small drumstick attached to the nucleus. Female cells contain two X chromosomes, but only one is actually used by the cell. Both Barr bodies and the drumsticks are thought to represent condensed non-functional X chromosomes in females. Cells from men lack both Barr bodies and drumsticks.

Female or not?

There was a distressing episode at the Olympic Games in the 1980s when a European woman who showed remarkable strength in field events was found to be genetically abnormal, lacking Barr bodies and drumsticks in her cell nuclei. Her strength was believed to be due to male characteristics. Sex tests involving Barr bodies are compulsory for certain athletics competitions these days.

Figure 2.5 shows the **karyotypes**, or complete sets of chromosomes arranged in pairs, for a normal woman and a normal man.

Sex linkage

Now that we have seen how sex chromosomes determine gender, let's return to Morgan and his flies. We can now explain his observations when he crossed his mutant white-eyed male with the normal red-eyed female. We will let **R** stand for the dominant allele for normal red eyes and **r** for the white-eye allele. This gene is located on the X chromosome. The Y chromosome in *Drosophila* carries no gene for eye colour. Figure 2.6 shows Morgan's crosses.

Morgan's first white-eyed male fly had an allele for white eyes on its X chromosome. We can represent this fly as $X^r Y$. The normal red-eyed female has the genotype $X^R X^R$. The sperm formed by the white-eyed male would be half X^r and half Y. All the eggs formed by the normal red-eyed female would be X^R.

What happens in the F_1 generation? Half of the F_1 offspring have the genotype $X^R X^r$. These flies are females that are heterozygous for eye colour. The other half of the offspring have the genotype $X^R Y$. These flies are males with a single gene for red eyes. So both the males and the females of the F_1 generation are red-eyed because the red-eye allele is dominant.

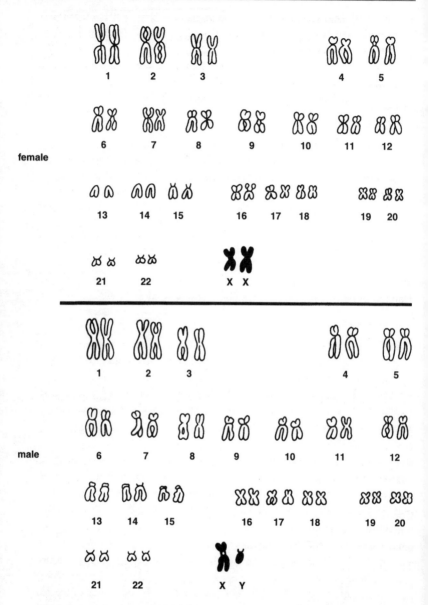

female

male

Figure 2.5 Female and male human karyotypes

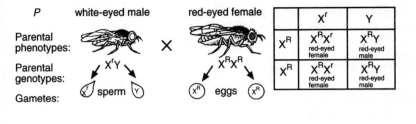

$F_1 \times F_1$

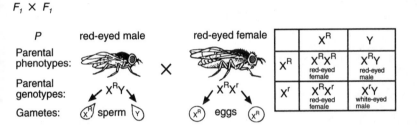

Figure 2.6 Sex linkage in fruit flies

Now look at the F_2 generation. These flies are offspring of an F_1 male and female. Half of the eggs produced by an F_1 female contain X^R, and half contain X^r. Half the F_1 sperm contain X^R and half contain Y. You can see that some offspring will be X^RX^R and some will be X^rX^R. These flies are red-eyed females. Of the male offspring, some will be X^RY, with red eyes; others will be X^rY, with white eyes. By the law of averages, about one quarter of the offspring will be of each type, so about three quarters will be red-eyed and one quarter white-eyed. All of the white-eyed offspring will be males. This is just what Morgan observed.

Could a white-eyed female appear in the next generation? Yes, if an X^RX^r female were mated with a white-eyed X^rY male. Can you work out the genotype of this white-eyed female?

The discovery of sex linkage brought a new and important principle to genetics. Sex-linked traits are not limited to *Drosophila*. In humans, for example, men are about ten times more likely to be red–green colour-blind than women. Also, haemophilia (the inability of blood to clot) is a sex-linked condition that affects far more men than women (see page 115).

The chromosome theory proved

Non-disjunction in fruit flies

As a graduate student Calvin B Bridges had helped Morgan in his discovery of sex-linked traits in *Drosophila*. In 1916 Bridges published a paper called *Non-disjunction as Proof of the Chromosome Theory of Heredity*. In this and later studies he provided evidence that genes are physically associated with chromosomes. He studied the way in which a gene for especially bright red eyes, called vermilion, was transmitted in *Drosophila*. Vermilion eyes were caused by a recessive allele which like the allele for white eyes seemed associated with the X chromosome. A cross between vermilion-eyed females and normal red-eyed males would usually give rise to normal red-eyed females and vermilion-eyed males, as shown in Figure 2.7.

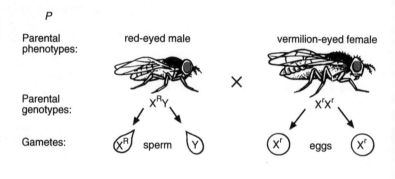

P

Parental phenotypes: red-eyed male   vermilion-eyed female

Parental genotypes: $X^R Y$ $X^r X^r$

Gametes: X^R sperm Y X^r eggs X^r

	Possible male gametes	
	X^R	Y
X^r	$X^R X^r$ red-eyed female	$X^r Y$ vermilion-eyed male
X^r	$X^R X^r$ red-eyed female	$X^r Y$ vermilion-eyed male

(Possible female gametes)

Figure 2.7 Normal inheritance of vermilion eyes in fruit flies

Very occasionally, no more than once in 2000 flies, the seemingly impossible happened. The cross produced a vermilion-eyed female! If the theory of recessive alleles were correct, this female must have inherited two recessive alleles for vermilion, yet she should have inherited the allele for red eyes from her normal red-eyed father. Where did the two recessive alleles come from?

Bridges deduced by a stroke of genius that the female parent's pair of X chromosomes failed to separate during the formation of her eggs. The normal method of egg production takes place by meiosis, as described on page 14, in which the normal number of chromosomes is halved. If the pair of X chromosomes did not separate they would be passed on as a pair to the daughter, along with the usual single sex chromosome from the father. By careful examination under the microscope, Bridges found these exceptional daughter flies to have three sex chromosomes instead of two, one Y and two X. The chromosome theory of heredity was finally established beyond all doubt. Bridges' cross is shown in Figure 2.8. **Non-disjunction** is the term used to describe this failure of the chromosome pair to separate during meiosis.

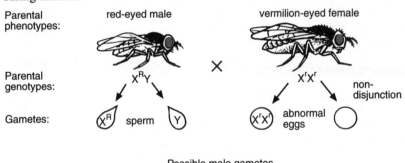

	Possible male gametes	
	X^R	Y
Possible female gametes: X^rX^r	$X^RX^rX^r$ red-eyed female usually dies	X^rX^rY vermilion-eyed female
0	X^R red-eyed male, sterile	Y lacks X chromosome and dies

Figure 2.8 Non-disjunction in fruit flies

Non-disjunction in humans

Klinefelter's syndrome and Turner's syndrome

The discovery of non-disjunction of chromosome pairs in sex cells was recognised as being of immense value in providing proof for the chromosome theory of inheritance, but was not at first considered to be of any practical value. Then, as often happens with a scientific breakthrough, this seemingly obscure scrap of knowledge was central in understanding a new discovery in human biology. Techniques were developed in the 1950s for studying human chromosomes, and it became possible to count them accurately and to identify the individual pairs. Early counts for humans gave a figure of 48, the same number as in our closest relatives the chimpanzee, gorilla and orang-utan. In 1956 a Javanese, Tjio and his Swedish colleague Levan showed that there were 46 chromosomes in human cells. This was confirmed by the two British scientists Ford and Hamerton from the Atomic Energy Research Establishment at Harwell.

Two interesting conditions had been recognised for some years by medical practitioners who thought that certain abnormalities of sexual development were associated with abnormalities of the sex chromosomes. The first group of patients were diagnosed as having **Klinefelter's syndrome**. They are apparently males but have underdeveloped testes and sparse facial and body hair. They often show some degree of mammary gland development and a feminine distribution of fat. These individuals are sterile, although capable of sexual intercourse. Many sufferers are first diagnosed only when they visit a clinic for the investigation of sterility. Klinefelter's patients often have learning difficulties.

The second group of patients were diagnosed as having **Turner's syndrome**. They are apparently females who fail to mature sexually, having undeveloped ovaries and also characteristic webbing of the neck and a low hairline. They often suffer deafness, dwarfism and malformations of the heart. These patients may also have learning difficulties.

In 1959 British scientists at Harwell and at Edinburgh were able to show that patients with Klinefelter's syndrome had not 46 but 47 chromosomes. These people have three sex chromosomes, XXY. Patients with Turner's syndrome, on the other hand, had not 46 but 45 chromosomes. They have only one sex chromosome, X, a genotype described as XO. How non-disjunction of the sex chromosomes could produce the two syndromes can be explained as in Figure 2.9, in the same way as Bridges had demonstrated in 1916 with the help of the humble fruit fly!

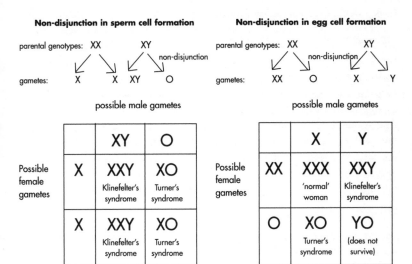

Figure 2.9 Non-disjunction in humans

Individuals with Klinefelter's syndrome can result from non-disjunction in sperm cell formation, leading to a sperm with both X and Y chromosomes. Fertilisation of a normal X-bearing egg cell with this sperm gives an XXY embryo. Alternatively, if non-disjunction occurred in egg formation, an egg with two X chromosomes would result which on fertilisation with a normal Y-bearing sperm would again give an XXY embryo, or with a normal X-bearing sperm would give an XXX embryo.

Individuals with Turner's syndrome can be produced as a result of non-disjunction in sperm cell formation, producing a sperm devoid of a sex chromosome. On fertilising a normal egg this would give an XO embryo. Non-disjunction during egg formation can give an egg without any sex chromosomes, and on fertilisation with a normal Y-bearing sperm this would give a YO embryo. However, it would be impossible for a YO individual to survive, as there is much important information on the X chromosome. You will see in Figure 2.9 another new genetic constitution, XXX. This has been found several times in women who show hardly any abnormalities.

It is interesting that although the sex chromosome abnormalities that occur in humans also occur in the fruit fly, the details of sex determination are different. The Y chromosome appears to have little function in the fruit fly. The XXY individual is a fertile female and the XO individual is a male, so that two X chromosomes produce a female and one X a male. In contrast the XXY human is a modified male and the XO human a modified female. In humans the Y chromosome appears to be the prime determinant of masculinity. This is confirmed by the discovery of people with Klinefelter's syndrome who have the genotype XXXY. These people are still masculine inspite of the three X chromosomes. Cells from a person with Turner's syndrome will have no Barr bodies, while those from a person with Klinefelter's syndrome will contain Barr bodies.

Down's syndrome

So far we have discussed non-disjunction of sex chromosomes. This failure of the chromosome pairs to separate during meiosis also takes place in autosomes. Such an example leads to **Down's syndrome**.

Down's syndrome is named after John Langdon Down who first described the condition in 1866, and wrongly suggested that the condition was due to a throwback to some ancient Mongolian ancestors. The condition is still sometimes known as 'mongolism' owing to the facial features associated with it, though the term is frowned upon as insensitive. The resemblance to the Oriental is at most superficial, and it is quite easy to recognise Down's syndrome in people of Oriental racial origin. Symptoms include narrow eyes with a persistent fold of skin over their inner edge and a slightly flattened face. People with Down's syndrome tend to be short and they suffer learning difficulties which range from mild to severe, but the children have a happy and affectionate disposition. Life expectancy used to be very short as sufferers succumbed to infections and problems related to heart defects. However with the use of antibiotics and heart surgery they live much longer today.

Like Klinefelter's syndrome, Down's syndrome is not uncommon and occurs about twice in 1000 live births in Western Europe and North America. Turner's syndrome, in contrast, is rare. Since Down's syndrome was first described, many clues have been found about what causes it. The syndrome usually occurs sporadically in an otherwise normal family and the incidence increases sharply as the age of the mother increases. The risk for a mother in her twenties is about one in 3000. For a woman of forty, however, the risk is about one in 100 and for women of forty-five and over the risk is as high as one in 50.

Since the early techniques of human chromosome analysis of the 1950s much research has taken place to find the cause of Down's syndrome. It is now known that the usual cause is non-disjunction of the small chromosome pair 21, giving rise to 47 chromosomes instead of 46. There are at least two other forms of chromosome abnormality which manifest themselves in Down's syndrome but these are much less common.

Why should the incidence of non-disjunction increase in older mothers? Several factors may be involved. The chromosomes pair up during the formation of egg cells in the ovaries before birth. However, they do not separate until the time of ovulation when the egg is released from the ovary, which may be anything from twelve to fifty years later. Chromosomes that remain paired for a longer time might separate less readily than those in contact for a shorter period. Also older women are more likely to have had greater exposure to ionising radiation, viruses and other agents known to increase the rate of non-disjunction.

Delayed fertilisation has also been suggested as a possible factor. Older women tend to have more delayed ovulations than do younger women, because of a changing hormonal balance. Also older women may have intercourse less frequently than younger women, so there is more chance of a delay between ovulation and fertilisation. There is only a period of a day or two after ovulation during which fertilisation can occur, but even a delay of this short time could be a factor. Most fetuses with an extra chromosome 21 are naturally aborted in all age groups, but perhaps more of them survive to birth in older women.

What about the older man – can he also be the source of the extra chromosome? The answer is yes, but not as frequently as the mother. Research carried out in the late 1970s suggested that men over the age of fifty-five show an increased chance of producing children with Down's syndrome. Overall there is evidence to show that non-disjunction in the father's sperm accounts for bout 25% of Down's syndrome cases.

Can heredity play a part in the chance of having a Down's syndrome child? If parents already have a child with Down's syndrome, are they likely to have another affected child in the future? These are two of the most common questions asked of genetic counsellors. Non-disjunction seems to be a chance occurrence, so the answer might seem to be a resounding no.

However, experience has shown that this is not always the case. Statistical studies suggest that for a woman in her thirties the chance is about one in 1500 if none of her previous children has the syndrome, but this increases to about one in 100 after a child with the syndrome has been born to her. Some factors must be operating to increase the risk in some women. There could be either internal physiological factors to do with the mother's body, or environmental agents that induce non-disjunction.

Linkage and chromosome maps

When Mendel formulated the principle of independent assortment, he was totally unaware of situations in which genes were not shuffled independently during the formation of the sex cells. Where independent assortment does not occur, this is often due to genes being linked together on a chromosome and so being passed on to the next generation together.

By the second decade of the twentieth century, scientists knew that the genes of the fruit fly were carried in four groups on four pairs of chromosomes. However, they did not know the positions of these genes in relation to each other. As early as 1911 the possibility had been considered of making maps of chromosomes to show the relative positions of genes on them. The crossing over of parts of chromosomes during meiosis described on page 14 was known as long ago as 1909. Morgan played with the idea that these crossings-over could enable him to deduce the order of genes in each chromosome. His logic was as follows:

Let us hypothesise that genes are arranged in a line along each chromosome, like beads on a necklace. Then during crossing over a gene may be transferred to another chromosome. If one chromosome carried the genes ABCDEFGHIJKL and its partner had the corresponding recessive genes abcdefghijkl, they could exchange parts so that the resulting chromosomes were ABCDEFghijkl and abcdefGHIJKL. This process is called **recombination**, because the genes in a pair of chromosomes are recombined during crossing over.

Now let us suppose that two genes are at opposite ends of a chromosome. When that chromosome breaks during crossing over, both genes may be interchanged, but it is most likely that only one will be interchanged and so they will be separated and appear different chromosomes. If, however, the genes are close together, separation is less likely. The chance that any two genes will be separated from each other during recombination is thus closely related to their distance apart on a chromosome. If this is true, Morgan suggested that it should be possible to plot the sequence of genes on a chromosome by analysing statistically how often any two of them were separated during recombination.

Besides Morgan, many research workers in this field attempted to test his hypothesis. They included his students Alfred H Sturtevant, Calvin B Bridges and Hermann Joseph Muller. All meticulously analysed the results of a multitude of matings between fruit flies which had particular mutations. Their patience paid off and they succeeded in placing these mutation sites in linear order. In 1913 Sturtevant published a research paper entitled *The linear arrangement of six sex-linked factors in Drosophila, as shown by their mode of association.* The first chromosome map was made in 1915. It showed the location of fifty genes on the four pairs of chromosomes of the fruit fly.

Since this pioneering work the number of mapped genes on the chromosomes of *Drosophila* has increased enormously. A single chromosome of this species is now known to carry over 500 genes. This was the start of gene mapping which has since been carried out on other species, including humans. There is more about this on page 121.

By 1915 the chromosome maps for fruit flies were described as statistical charts rather than maps showing the positions of actual genes. Using these charts geneticists could predict the distribution of characters in the offspring of any particular cross. Viewing real genes was thought to be impossible because of their small size, but in 1930 D Kostoff unearthed the work of Balbiani from 1881 (see page 45), who first described giant chromosomes from the salivary glands of fruit flies. When these are stained and squashed on a microscope slide, thousands of crossbands are visible along their length. Kostoff suggested that these bands might correspond with the arrangement of the genes. Then in 1933 Theophilus Shickel Painter of the University of Texas wrote the following account after intensive studies of giant chromosomes:

> The same chromosomes, or characteristic pairs thereof, may easily be recognised in different cells … . This discovery places in our hands, for the first time, a qualitative method of chromosome analysis … . By studying chromosome arrangements of known genetic character, we can … construct chromosome maps with far greater exactness than had been heretofore possible.

Detailed analysis combined with endless patience went into comparing the changes in appearance of these chromosomes with the changes in appearance of the fruit flies that possessed them. These physical maps did not look like the statistical ones, although the order of the genes was the same, as Figure 2.10 shows. One reason for the difference was that some parts of the chromosome are more likely to cross over than others, and so are farther away from each other on the statistical map than they really are.

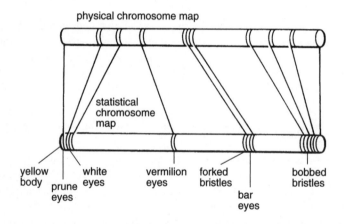

Figure 2.10 Statistical and physical chromosome maps do not match exactly because the likelihood of crossing over is not the same everywhere along the chromosome

Mapping chromosomes demonstrated beyond any shadow of doubt that genes are arranged in a line along chromosomes, just like beads on a necklace. The next question was: how big is a gene? It was answered in 1934 by one of Morgan's ex-students H J Muller, and A A Prokofyeva. They used recombination studies to show that the maximum length of a gene from a fruit fly was 1250 nanometres (one nanometre, nm = 0.000 000 1 cm).

More mechanisms of inheritance

As we have seen, Mendel's classical F_1 generation ratios 3 : 1 (for a pair of contrasting characters) or 9 : 3 : 3 : 1 (for two pairs of contrasting characters) are not seen in the offspring of all matings. Codominance and various forms of gene linkage lead to exceptions to Mendel's laws of genetics. It did not take the classical geneticists of the twentieth century long to fathom out how these two complications evaded Mendel's neat and elegant predictions. As time went on, other mechanisms were unravelled for inherited characters that seemed to defy the principles of segregation and independent assortment. Unusual genetic ratios were interpreted in many ways. Some of the more common anomalies are explained below.

Lethal alleles

In 1905 the French scientist Lucien Cuenot was carrying out genetic crosses with mice. He crossed a mutant mouse with yellow fur with a pure-breeding (homozygous) normal grey type (**yy**). The result was a 1 : 1 ratio of yellow to grey. He concluded that the yellow mice were heterozygous (**Yy**) and that yellow was dominant. When two yellow offspring were mated together, they produced young in the ratio of 2 yellow : 1 grey. This was a strange departure from the expected classical 3 : 1 ratio. Cuenot hypothesised that the allele for yellow must be **lethal** in the homozygous form **YY**. Female **Yy** individuals were dissected to test his hypothesis and aborted fetuses were found in them, thus confirming his prediction. The modified 2 : 1 ratio was thus explained as shown in Figure 2.11.

A common lethal allele in plants causes lack of chlorophyll. These alleles are usually recessive. Homozygous recessives are unable to carry out photosynthesis at an early stage of seedling development and die. It is relatively easy to show the effects of such lethal genes in mice and plants but in humans the situation is quite different.

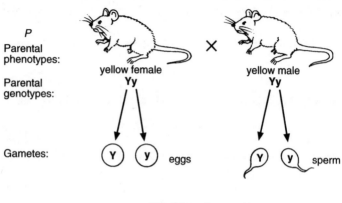

Possible male gametes

×	Y	y
Y	**YY** lethal	**Yy** yellow
y	**Yy** yellow	**yy** grey

Possible female gametes

Figure 2.11 Effects of a lethal allele in mice

The extreme complexity of the human body is such that hundreds of thousands of genes are needed to code for a single life. A mutation of any one gene could be lethal, and it is estimated that each person carries two or three recessive lethal alleles. There are so many different kinds of lethal allele that the chances of having a child with a partner who has even one lethal allele that matches one of yours is very remote. Should such a matching occur, the lethal gene would be expressed as a double recessive allele in only one in four of your children in any case. However, the chance of both parents carrying the same lethal gene increases when they are close relatives. People with a common ancestry are more likely to have many genes in common than non-related individuals. As a result stillbirths, spontaneous abortions and deaths just after birth are much more common in infants who are the results of incest than in children of unrelated parents.

Lethal genes can act at different times during the development of an individual. Some genes prevent complete cell division (mitosis) during the earliest stage of development of the fertilised egg. Others cause the embryo to die before it reaches the uterus after fertilisation in the oviduct. Some genes interfere with implantation, in which case the embryo is expelled from the uterus. All of these genes would have their effects so early in the development of the embryo that the woman would never know that fertilisation had occurred. A lethal gene that prevents the normal formation of the heart shows its effect about three weeks after conception, at the time when the embryo's circulatory system becomes essential for its existence. A gene that interferes with normal kidney development becomes evident within a day or so of birth. Late-acting lethal genes are rare but include Tay-Sachs disease, a type of muscular dystrophy which kills several years after birth, and Huntington's chorea (see page 36) which usually causes death at about forty to fifty years of age.

Gene interaction

During the first decade of the twentieth century Bateson and Punnet came across some unexpected results after mating chickens with various comb shapes. Two different forms of comb are called pea and rose. Both of these are true breeding but when crossed together produce a hybrid with an entirely different shaped comb called walnut. If these walnut types are crossed, the next generation has four comb types: walnut, rose, pea and another new type called single, in the ratio of 9 : 3 : 3 : 1. Figure 2.12 shows the cross.

Bateson and Punnet solved the puzzle of this cross by suggesting two independent genes which they called **P** for pea and **R** for rose. Pea and rose are both dominant to single. A chicken that has the dominant allele of the gene **P** has a pea comb, and similarly a chicken that has **R** has a rose comb. A chicken without any dominant alleles (**rrpp**) has a single comb caused by interaction between the two homozygous recessive pairs of alleles. When the dominant alleles of both **P** and **R** are present together, they also interact to give the walnut comb. Pure-breeding chickens with pea combs therefore have the genotype **PPrr**, and those with rose combs are **RRpp**. When these two types are mated they produce chicks with the genotypes **PpRr**. When these chicks grow into adults and are mated together they produce a 9 : 3 : 3 : 1 ratio, but it differs from Mendel's cross with two pairs of contrasted characters because there is only one

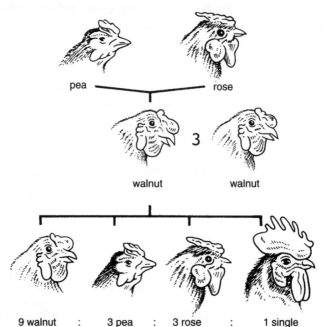

9 walnut   :   3 pea   :   3 rose   :   1 single

Figure 2.12 The inheritance of comb shapes in chickens

character involved – the comb shape. The offspring do not show just new combinations of characters already present in the parents but novel forms of this one character due to interactions between two controlling genes. The genotypes are shown in Figure 2.13.

In this example the only chickens whose genotypes you can be sure about just by looking at the shape of the comb are those with single combs (**pprr**). Most mass-produced battery hens have a single comb and are also albino, another homozygous recessive character. It is likely that such chickens will also have other known recessive genes. The reason for this is that the poultry farmer must be certain of the genotype of the flock to be able to guarantee the quality of the product and also to predict growth rates and egg production. Farmers cannot afford to introduce unwanted genes into their flocks by mating with unknown genotypes, and double recessive characters leave no doubt as to the genotype.

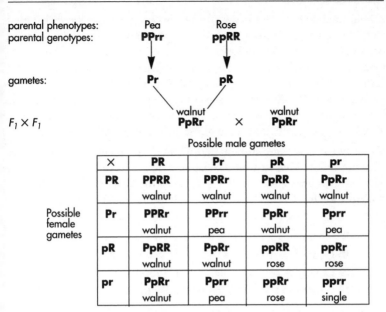

9 walnut : 3 pea : 3 rose : 1 single

Figure 2.13 The interaction of two genes controlling comb shapes

The inheritance of blood groups

In 1900 Karl Landsteiner of Vienna was a pioneering specialist in the study of human blood. He made an important discovery while investigating the problems of mixing blood from different people. This research led to the understanding of blood grouping, and saved the lives of vast numbers of people through blood transfusion during the twentieth century.

Landsteiner found that mixing blood from certain people resulted in clumping together of the red blood cells. However, this clumping together did not always take place. Why did it happen in some cases but not in others?

As far back as William Harvey's discovery of blood circulation in the seventeenth century, ancient quacks attempted blood transfusions between animals and humans. Some transfusions between humans were also tried and indeed some were successful. Landsteiner's discovery changed desperate hit-or-miss techniques into one of the surest aids known to medicine. He discovered that the red blood cells from some people have a certain chemical on the cell surfaces, while the red blood cells from other people do not. The chemical may be of two types – A or B. Some red cells have only one type A or B, some have both A and B and some have neither. Today we can classify all human blood as type A, B, AB or O.

These blood types are an inherited feature. The method of inheritance does not follow Mendel's laws because there are three alleles rather than a single pair that code for blood types – the feature is controlled by **multiple alleles**. Of the three possible alleles for blood type, only two are present in any individual. The alleles are represented as I^A, I^B and I^O:

- Genotypes $I^A I^A$ or $I^A I^O$ produce type A blood.
- Genotypes $I^B I^B$ or $I^B I^O$ produce type B blood.
- Genotype $I^O I^O$ produces type O blood.
- Genotype $I^A I^B$ produced type AB blood.

The inheritance patterns of these alleles are illustrated in Figure 2.14.

An understanding of the inheritance of blood groups may be used in settling a paternity suit. A woman may allege that a particular man is the father of her child. The woman's and man's blood can be tested and the blood groups may help to disprove the allegation.

Consider a father with blood group O and a mother with blood group A. The possible number of blood groups of the children is then quite small:

Possible genotypes of the parents:	Possible blood groups of the child:
mother $I^A I^A$ or $I^A I^O$	A or O
father $I^O I^O$	

If the child has A or O type blood, the man may indeed be the father. If the child's blood type is B or AB, this man could not possibly be the father. Remember though that this only gives proof that the person *could not* be the father. It is impossible to tell if he *is* the father of the child using only blood group inheritance. Genetic profiling is a surer method of proving such cases (see page 136).

Since Landsteiner's discovery of the four main blood groups, many other sub-groups have been found. Arguably the most important of these was

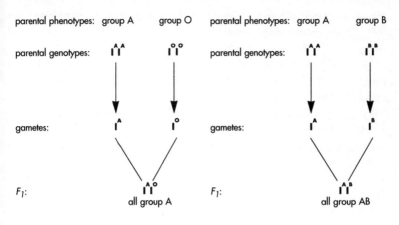

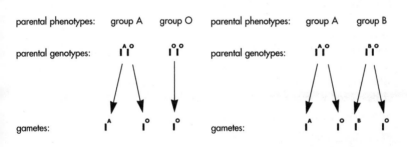

Figure 2.14 The inheritance of blood groups

again discovered by Landsteiner shortly before the Second World War in 1939. This was the rhesus factor.

Landsteiner and his co-worker Weiner injected blood cells from a rhesus monkey into rabbits and guinea pigs. These subsequently produced a protein called an antibody in their blood which reacted with the rhesus monkey's cells. Later this antibody was shown to cause clumping together of the red blood cells of about 85% of humans. Landsteiner and Weiner suggested that the antibody was reacting with a previously undescribed chemical which may or may not be present on human red blood cells. They called this chemical the rhesus antigen. People with the antigen are called rhesus positive (Rh+) and those without it are rhesus negative (Rh–). The discovery in 1941 by Levine that the presence of both the antibody and the antigen could cause clumping together during blood transfusions established the medical importance of this blood grouping.

When the inheritance pattern of the rhesus antigen was studied, it became clear that it is a dominant factor and follows Mendel's inheritance laws. The consequences of this can cause problems – if the father is Rh+ and the mother is Rh– the child could well be Rh+. If there is any seepage of blood across the placenta during pregnancy, the mother's blood could react with the child's blood and cause clumping together of the red blood cells. There are techniques available to counteract this problem, so that after initial treatment both mother and child can remain healthy. As long as the blood groups of the father, mother and child are known in advance then there are ways of preventing harm to the baby or the mother.

The inheritance of eye colour

It is commonly thought (and commonly taught!) that eye colour is inherited in a simple dominant/recessive Mendelian manner. Brown eyes are considered to be dominant to blue eyes. The real pattern of eye colour inheritance, however, is considerably more complicated than this.

Eye colour is determined by the amount of a brown pigment known as melanin that is present in the iris of the eye. If there is a large quantity of melanin present on the front surface of the iris, the eyes are dark. Black eyes have a greater quantity of melanin than brown eyes. Without a large amount of melanin on the front surface of the iris, the eyes will appear as blue, not because of a blue pigment but because blue light is reflected from the iris. The iris appears blue for the same reason that deep bodies of water tend to appear blue – there is no blue pigment in the water, but blue wavelengths of light are returned to the eye from the water. People appear to have blue eyes because the blue wavelengths of light are reflected from the iris.

Colours such as green, grey and hazel are also produced by the varying amounts of melanin in the iris. If a very small amount of brown melanin is present in the iris, the eye appears green, whereas slightly larger amounts of melanin produce hazel eyes.

Several different genes are probably involved in determining the quantity and placement of the melanin and therefore in determining eye colour. These genes interact to produce a wide range of eye colour. Two more observations on eye colour further complicate a seemingly simple inheritance pattern:

- Albinos have no pigment at all in the iris and so the blood vessels in the iris show up as pink eye colour. This is commonly seen in white mice or white rabbits. In humans the lack of pigment in the iris gives no protection against strong sunlight, though dark glasses can be worn as a substitute. The albino gene is inherited as a recessive character.
- Newborn babies all have blue eyes that may later become brown or green. At the time of birth they have not yet begun to produce melanin in their irises.

The inheritance of skin colour

The inheritance of varying amounts of melanin in the skin is also controlled by the interaction of many genes – an example of **polygenic inheritance**. Skin colour is determined by genes at several different places on chromosomes. A number of different pairs of alleles combine their effects to determine the characteristics. According to some human geneticists, genes for skin colour are located at a minimum of three different places on chromosomes, or **loci** (singular locus). At each locus the allele for dark skin is dominant to the allele for light skin. Therefore a wide variety of skin colours is possible depending on how many dark-skin alleles are present.

The inheritance of height

Polygenic inheritance is very common in determining such characteristics that show a range, rather than being either one thing or the other. For example, people show great variation in height – there are not just tall and short people like Mendel's pea plants; there is a wide range showing every possibility between some adults who are as short as one metre and others who are taller than two metres. By plotting a histogram as shown in Figure 2.15, we see **continuous variation** between the limits within which most individuals fall. We can also see how the more extreme cases are spread on each side. In making such measurements, however, we must try to make sure that all other factors are kept constant – for example, when measuring height in a population of humans the sample must all be of the same age.

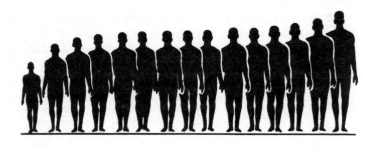

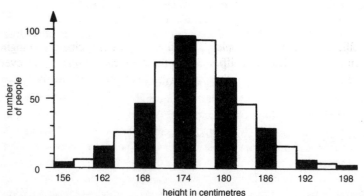

Figure 2.15 Height in humans shows continuous variation – many different heights are possible

In contrast to such polygenic features, characters which are coded for by single dominant or recessive alleles produce histograms which show **discontinuous variation**. An example is the height of pea plants, shown in Figure 2.16.

Discontinuous variation sorts individuals into distinct groups with no overlapping.

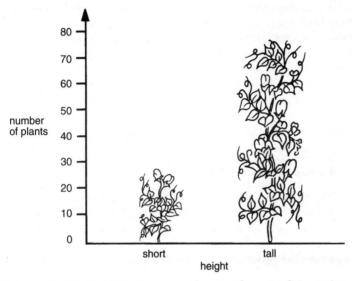

Figure 2.16 Height in pea plants shows discontinuous variation – all plants are either tall or short

Intelligence is another example of a continuously varying character, ranging from those with learning difficulties to those we call geniuses with every level in between. Many of these polygenic traits may also be influenced by outside environmental factors, such as diet, disease, accidents and social factors.

The role of the environment

As geneticists try to understand what has influenced an organism's heredity, patiently obtaining masses of data in order to form an hypothesis, they must not forget that all organisms live in complex environments, and that non-genetic factors such as diet may have just as great an influence as genes on some characteristics. Heredity determines what an organism may become, not what it *will* become. A plant may inherit genes for tallness but may not grow tall in poor soil. An organism's phenotype or outward appearance depends on a combination of inherited characters and the effects of the environment.

A simple example can illustrate this. Imagine a litter of Alsatian puppies – they all have genes from the same parents. If we follow and compare the growth of these puppies in their separate homes, we might see that one out of the litter becomes less strong and not so large as the rest. The reason could be its genes, or it could be that it has not been given the correct diet and exercise. In other words, the effects of the environment may well have influenced its phenotype. By altering the environment we can alter the phenotype. On the other hand, a litter of corgis all kept together will grow up to be smaller dogs than the Alsatians no matter how well fed they are. Any small difference in size is then likely to be genetically determined. The genes determine the phenotype if the environment is kept the same.

What is the environment?

The **environment** of an organism is not just the place where it lives – the term includes all the external conditions that govern its life or growth. These external conditions encompass the nutrients and any other materials taken in the organism, climate and any physical forces acting on the organism from the outside.

The effect of diet

Figure 2.17 shows the effects of improved diet on the growth of some Japanese boys. The groups of boys were brought up in different environments – some were born and raised in the USA, where a better diet was available; the others were brought up in Japan. Between 1900 and 1952 the diet and standard of living in Japan improved markedly. From the graphs you can see that the American-born boys were the tallest at all ages. The improved diet and standard of living in Japan resulted in an increase in height between 1900 and 1952. The same has happened in Britain – children are now taller and heavier, age for age, than they were 60 years ago.

Another example of the effect of diet on phenotype are the abnormal heights and body shapes of people suffering from insufficient levels of vitamins or minerals in the diet, causing deficiency diseases such as rickets which

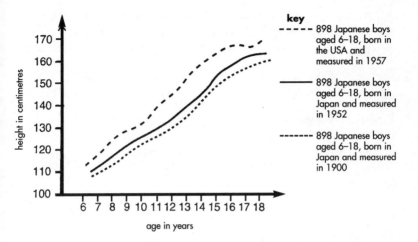

Figure 2.17 The effect of improved diet on height illustrates the role of the environment in determining the phenotype

is due to lack of vitamin D or lack of calcium. People with rickets have soft bones which become distorted under the weight of the body.

The effect of temperature

The temperature of the environment can affect the phenotype of an organism. Some early investigations showed that the fur on certain small mammals was black on the areas of the body which tend to be coolest. For example, the Himalayan rabbit is pure white apart from its ears, paws, nose and tail which are black. The distribution of black fur on many types of cats is also related to body temperature.

Temperature also affects at least one trait in the wing shape of the fruit fly. Some flies have wings that curl up sharply. If these flies are raised at a temperature of 25 °C, the wings curl. If the same strain of flies are raised at 16 °C, the trait appears very rarely. This is illustrated in Figure 2.18. The gene for curly wings is still present in the apparently normal flies raised at 16 °C as the next generation shows curly wings if the temperature is suitably high.

The effect of light

A factor that affects the appearance of plants is light. Most plants inherit the ability to produce chlorophyll, the green pigment that allows plants to make their own food by photosynthesis. However, light is also essential for the

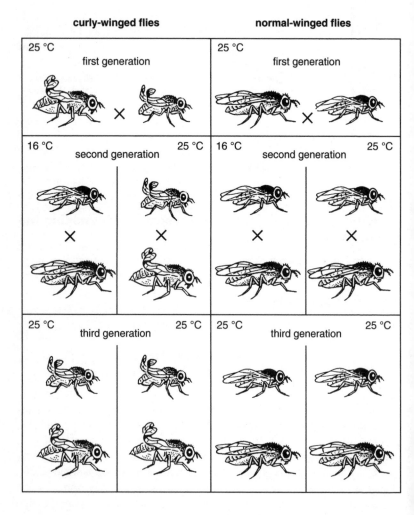

Figure 2.18 The effect of temperature on phenotype in fruit flies

production of chlorophyll. Seedlings grown in the dark lack the normal green colouring, and do not begin to make chlorophyll and turn green until they are exposed to light. Only under suitable environmental conditions can we find whether these plants have the genetic potential to make chlorophyll.

Some parasitic plants can exist without chlorophyll because they obtain their nutrients from their host. These plants do not have genes for chlorophyll production and will not develop it even when exposed to light. Dodder is such a parasite, a plant with long yellow stems that coil around the host plant such as nettle or clover and produce outgrowths that penetrate the host plant's stem and collect nutrients from it.

Variation and mutation

If you looked casually at a group of animals of one species, you might think that they were all alike. However, anyone who has to care for domesticated animals such as herds of cattle soon learns to distinguish between them. Plants can vary just as much as animals. No two individuals are exactly alike – even identical twins have different fingerprints. This difference between members of the same species is called **variation**, and it has both genetic and environmental causes.

A girl tends to resemble her sister or her mother more closely than more distant relatives, and she resembles her relatives more closely than her unrelated friends. These resemblances are due mainly to genes. The resemblance of identical twins is due totally to heredity, and the differences that enable us to tell one twin from another are due to environment.

Although sisters have the same parents, they do not have exactly the same genes. The separation of the chromosomes in meiosis (see page 14) and the random way in which they recombine produces variation. Even the simplest organisms have hundreds of genes, and complex organisms have thousands. A tremendous amount of variation is possible from different combinations of this number of genes. Furthermore, crossing over of parts of chromosomes during meiosis increases variation by making new combinations of genes. The mixing of these new combinations at fertilisation ensures that all individuals are unique. New potential for variation comes about when there is a change in the genes themselves – a **mutation**. A mutation is a change in the chemical structure of one or more genes, which can result from a mistake when the genes are duplicated during cell division. Most mutations are recessive and are masked by dominant normal genes, so except for some sex-linked genes (see page 42) mutations do not usually show up until two of the same mutant genes occur in the same individual, making the individual homozygous for the mutant gene.

Some mutations produce major changes; other cause only minor changes in body chemistry. If the change is highly unfavourable for the organism, the mutation will probably be lost from the species because the affected organism is unlikely to survive and have offspring which will inherit the mutant gene. Sooner or later, however, the mutation will occur again.

It is important to realise that mutations can be favourable as well as unfavourable, and are a valuable source of variation. Although we may speak of mutant genes and normal genes, the genes we now call normal were once mutations. Because they were favourable, they have been passed from generation to generation and have become part of the normal collection of genes. Mutation is all part of the process of evolution.

Radiation in the environment

Mutation and radiation

All living things are constantly exposed to a certain amount of radiation. The sources of this background radiation are shown in Figure 2.19 – the largest natural source is radioactive materials in the Earth's crust, along with cosmic radiation from outer space. This background radiation is in part responsible for the mutations that occur in all organisms.

Radiation is a series of subatomic particles or high-energy rays. When radiation hits an atom it can strip off an electron and release a large amount of energy, so any biological molecule hit by radiation may be destroyed. This in itself is enough to cause considerable damage to a living cell.

It has been known for a long time that an increase in radiation levels causes an increase in the mutation rate – the rate at which genes are altered. This knowledge has been used in genetic experiments to produce mutations in experimental organisms. Radiation treatment has produced some useful plant varieties. It has also been known that even very low levels of radiation have an effect on the mutation rate. These findings have real significance for us in the modern nuclear and technological age.

The human race has increased radiation levels in several ways. We use X-rays for medical purposes, and we test nuclear weapons. The dumping of

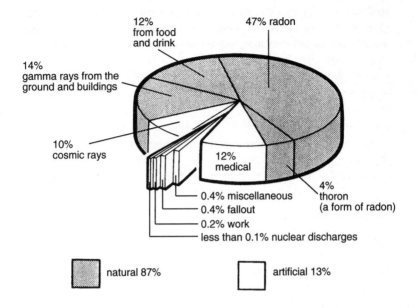

Figure 2.19 Sources of radiation in the environment

radioactive waste from nuclear power stations is a problem which has not been fully solved with modern technology. This increase in radiation may be only a fraction of the naturally occurring background radiation, but it must be emphasised that any increase in radiation levels will increase the mutation rate.

Radiation has its greatest effect on the bone marrow and liver cells, which are very active tissues in the body, and also on the sex organs where it may cause mutations which will be passed on to the next generation.

Radiation is a known cause of anaemia and also leukaemia, a cancer which affects bone marrow cells. In leukaemia, the mechanisms that control the manufacture of blood cells break down leading to an imbalance of red cells, white cells and platelets in the blood.

Children who live close to river estuaries flanked by heavy industry have unusually high levels of radioactivity in their tissues. Radiation is present in petrochemical pollution and the tidal action in estuaries concentrates the pollutants. Nuclear waste leaked from the Sellafield reprocessing plant in Cumbria has been detected in the Irish Sea where fish are monitored for radioactive contamination. Grass and seaweed samples near nuclear power stations are also analysed to check that radiation levels are not too high. Strontium-90 and caesium-137 have attracted considerable attention because they are radioactive substances that readily accumulate in living organisms. For example, fish caught in the North Sea have been found with caesium-137 in their flesh. Strontium-90 is similar to calcium and becomes concentrated in bones.

X-rays and gamma rays have great penetrating power, and great care must be taken to minimise exposure when X-rays are used. Again the most active tissues such as developing embryos are most sensitive to their effects, and that is why pregnant women are not X-rayed.

Radioactive waste

The production of highly radioactive waste is a problem with all technologies that use radioactive materials, including nuclear power stations. The waste must be disposed of safely in places from where it cannot be recovered. Solid high-level waste may be disposed of on the surface of the deep ocean floor, under the surface of the ocean floor or under the surface of the land. Whichever method is used, there will always be possible routes by which radiation may return to the environment. Some of these are illustrated in Figure 2.20.

The legacy of some nuclear disasters – Dounreay

A massive explosion shook the Dounreay nuclear plant on the coast of northern Scotland in the early morning of 10 May 1977. Two kilograms of potassium and sodium had been deposited in a shaft containing radioactive waste in which there was sea water. The subsequent uncontrolled release of energy was so violent that a huge concrete lid was blown off and its steel top-plate was thrown 12 metres to one side. Five-tonne concrete blocks at the mouth of the shaft were badly damaged and scaffolding 40 metres away was distorted. An eyewitness described a plume of white smoke blowing out to sea, but it was not until 1995 that it was revealed that the ground around the shaft had been littered with radioactive particles.

Since the accident up to 150 such particles have been found on the beaches at Dounreay. It turned out to be the most serious accident ever to have happened there and arguably resulted in the most dangerous release of radioactive particles ever in the history of the British nuclear power

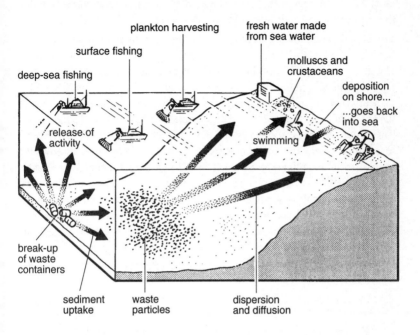

deep-sea fishing

surface fishing

plankton harvesting

fresh water made from sea water

molluscs and crustaceans

deposition on shore...

...goes back into sea

release of activity

swimming

break-up of waste containers

sediment uptake

waste particles

dispersion and diffusion

Figure 2.20 Routes by which radiation could return to the environment from radioactive waste

industry. By 1987 investigations had begun into a cluster of childhood leukaemia cases around Dounreay. A link between the radioactive particles and the number of cases of leukaemia was suspected. However, it is interesting to compare the relatively limited media coverage of this incident with the enormous international mass press, radio and television space devoted to Chernobyl nine years after the Dounreay incident.

The nuclear disaster at Chernobyl

The first warning of the Chernobyl incident came at 9.00 a.m. on Monday 28 April 1986. Technicians at a nuclear plant 60 miles north of Stockholm, Sweden, began to see alarming blips across their monitors. This was the first sign of high levels of radiation. At first they assumed that there was a leak at their own nuclear plant so they immediately searched for the cause. They found nothing, but radiation levels continued to rise in the area. After concluding that the radiation leak was not attributable to them, they concentrated their attention on the south of the Soviet Union.

Winds had been blowing from the south into Scandinavia for several days, but when the Swedish authorities first requested information from the Soviet Union they were told nothing. Later that night it was announced that an incident had occurred – there had been a nuclear accident at the Chernobyl nuclear power plant 80 miles north of Kiev. Satellite photographs revealed the extent of the disaster, showing that the plant had experienced a meltdown. This occurs when the cooling system fails and the radioactive core overheats to melting point. The molten radioactive material may burn through the ground and contaminate the water table. When Soviet technicians flooded the area with water, a reaction occurred between steam, uranium and graphite, producing a flammable gas which blew up the reactor. A cloud of radioactive material was released high into the atmosphere. The winds blew this cloud across Poland and Scandinavia, showering these countries with radioactive chemicals. The direct risks to health were inhalation and skin irradiation, and because of indirect risks the consumption of fresh vegetables and milk produced in several countries was restricted.

By March 1989 the effects of this disaster were becoming obvious in farms just outside the 30-kilometre exclusion zone around Chernobyl. Deformed animals were born in alarming numbers compared with the five years prior to the accident. Some animals lacked heads, limbs, eyes or ribs. Most pigs had deformed skulls. Radiation levels on the farms were 148 times as high as background level. Women from the area were advised not to have children and the incidence of cancer cases doubled.

By 1995 research into the wildlife of the area revealed the extent of mutations caused by the radiation. Studies of small mammals showed the full effect on animal populations and the potential problems for humans. No fewer than forty-six mutations in just one gene were found in nine voles collected from the 30-kilometre exclusion zone around Chernobyl. There were just four mutations of the same gene in samples taken from outside the zone. The study raised questions about how many mutations a population can tolerate without dying off.

The effects of radiation on mutation also have implications for humans. The Second World War atomic bomb survivors of Hiroshima and Nagasaki were recorded as having an increased rate of stillbirths and cancer but techniques in gene analysis were not then available. Modern researchers are in no doubt that genetic changes were induced by radiation from the bombs.

3 | THE NATURE OF THE GENE

What is a gene?

How big is a gene?

By the mid-1930s it was becoming obvious to the increasing number of research geneticists that the concept of the gene was complex. Various definitions of a gene were suggested depending on the particular line of research of an individual scientist. These definitions were interpreted to mean different things. For example, the size of a gene as defined from recombination studies (see page 56) seemed to be about 100 nanometres. However, by using X-rays to produce mutations and calculating the area targeted by the X-ray beam, scientists concluded that irradiating an area of only 10 nanometres in diameter could be enough to cause a mutation. Should the target area be considered to be a complete gene?

Just prior to the Second World War the American geneticist H J Muller carried out pioneering studies on mutation rates using irradiation. Working with *Drosophila*, he found that a sudden burst of X-rays to parent flies increased the mutation rate in the offspring by 150 times. Soon governments became interested in this work because of the potential military implications. A German refugee Charlotte Auerbach working in Edinburgh was commissioned to work on chemicals that caused mutations. The horrific effects of the phosgene or mustard gas used in the First World War seemed like a good starting point. This gas produced appalling burns on the skin which took months to heal – symptoms which were very similar to radiation effects. Her research confirmed that phosgene causes mutations, but this was classified information until after the Second World War. So the gene was something that could be damaged by chemicals and by X-rays. Increasing the level of X-radiation proportionally increased the chances of causing mutation, so it was concluded that the gene must be a chemical capable of being altered by X-rays.

The position effect

The discovery followed that the effectiveness of a gene in expressing itself depended on its relation to its neighbouring genes. This was called the position effect, and was studied in detail by Muller. He suggested that the effects of a gene might partly depend on overlapping regions between neighbouring genes. The exact limits of an individual gene would therefore depend upon its neighbour and would constantly shift at each recombination during meiosis. He and his colleague Daniel Raffel hypothesised in 1940 that the gene might be divided into still smaller pieces which under some conditions could act as independent units, for example, in recombination. The hypothesis was followed up by Guido Pontecorvo, working at the University of Glasgow. In 1952, while looking for subdivisions of the gene, he pointed out that mutations at several sites close together might produce different effects. This depended on whether the sites were all on one partner of a chromosome pair or whether they were distributed between both partners. He used the fungus *Aspergillus nidulans* for his studies. This fungus can synthesise a chemical called biotin, but its ability to do so can be destroyed by a recessive mutation at any one of three closely grouped sites on the chromosome. If there were two mutations on the same chromosome of a chromosome pair, their effect was masked by the other normal chromosome carrying the dominant factor. On the other hand, if there was a mutation at one site on one member of a chromosome pair and another at another site on its partner, the fungus died because it could not produce biotin.

The interpretation of these observations pointed to one of two possibilities. Either three separate genes were involved or there were three mutant sites within one gene. Pontecorvo considered the second possibility to be the more likely, which meant that the gene could be separated into three constituent parts. Thus the presence of any one of the three mutations on one chromosome and of another on the other chromosome would put the gene out of action on both. In suggesting this explanation Pontecorvo was leading the way in the detailed genetic analysis of sites within a single gene. Since then work in this field has expanded greatly, culminating in the attempt to determine the position of all the human genes (see page 121).

Separating the gene

Because neighbouring sites within a gene are so close together, it is difficult to plot mutation sites within a gene. The closer together two sites are on the chromosome the smaller the chance of the sites being separated during recombination. This means that mutation sites within a gene are only rarely separated, so it is necessary to search through millions of individuals to find one in which site separation has occurred during recombination. Even fruit flies are of no use here because they produce at most a few thousand individuals and each generation takes about twelve days. Simpler organisms are needed for this role, that breed by the million and within minutes rather than days. Suitable candidates for such a role include only bacteria or viruses which infect bacteria (bacteriophages). Test tube experiments using such microbes can produce the necessary data within twenty minutes.

A bacteriophage called T4 was chosen for some classic and meticulous investigations by the American Seymour Benzer. He used a mutant form of the T4 virus called rII. When the virus is mixed with two different strains of the bacterium *Escherichia coli*, the B and K strains, rII mutants can infect the B strain but not the K strain.

Benzer found that if he mixed two different rII mutants together with *E. coli* K, some of the viruses would infect the K strain. The genetic material from the two viruses had recombined while they were together in the bacterium. The genetic material of the viruses could recombine to form a normal type virus, but this was true only for certain pairs of rII mutants. By fine mapping techniques, Benzer located the mutation sites on these mutants. He found that the sites always lay towards opposite ends of a region of the genetic map of the rII virus. He divided this region into two parts, which he called cistrons, designating one A and the other B.

If an rII mutant with a mutation in the A cistron was recombined with another bearing a mutation in the B cistron, it was possible for a virus made up of a normal B cistron plus a normal A cistron to result. Conversely, if both mutants had their mutant sites in the same cistron, no amount of recombination could make the virus with both cistrons functioning correctly. Benzer had succeeded in dividing the rII gene into two clear-cut parts, each of which could function independently of the other.

It had taken almost a hundred years, from the 1860s to the 1950s, to get somewhere close to answering the question: 'What is a gene?'. However, the chemical structure of the gene and an explanation of how it works were still unknown in 1950. A combination of techniques in genetics and molecular biology were needed to unravel this mystery.

The chemical structure of the gene

The analysis of nucleic acid

In 1869 a young Swiss biochemist named Friedrich Miescher discovered a chemical which proved to be the most significant substance in the structure of the gene. He named the substance nuclein because it existed only in the nuclei of cells. He needed lots of cells for his studies and the source of his research material was pus from discarded surgical bandages obtained from a local hospital, the pus providing a plentiful supply of white blood cells from which he could obtain the nuclei. This research was not a job that many would envy but by persevering with this bizarre supply of cell nuclei he discovered a fundamental group of biological chemicals called nucleic acids.

Miescher eventually changed his source of material to salmon sperm. Each spring he visited the Rhine falls above his home town of Basel when the salmon were leaping the falls. With the help of local fishermen he obtained enough nuclear material to last him through the following year. Miescher soaked the sperm in strong salt solution and precipitated the strands of nucleic acid by adding water. This had to be carried out at low temperatures so in those days before refrigerators the only way around the problem was to work in an unheated room in the middle of winter. His methods were described in a letter to a friend:

> When nucleic acid is to be prepared, I go at five o'clock in the morning to the laboratory … . No solution can stand for more than five minutes, no precipitate more than one hour being placed under absolute alcohol. Often it goes on until late in the night. Only in this way do I finally get products of constant phosphorus proportion.

Working under these rigorous conditions, Miescher succeeded in analysing nucleic acid to find the chemical elements present. Besides phosphorus, there were carbon, oxygen, hydrogen and nitrogen. Other scientists continued the research into nucleic acid. The German Albrecht Kossel discovered that it contained combinations of atoms built around two basic carbon–nitrogen configurations called **purine** and **pyrimidine** rings, shown in Figure 3.1.

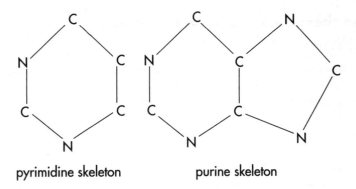

pyrimidine skeleton   purine skeleton

Figure 3.1 Two types of rings in the bases found in nucleic acid

Kossel isolated two pyrimidines, which he named cytosine and thymine, and two purines, which he named adenine and guanine. For this he received the Nobel prize for medicine in 1910. Another pyrimidine, uracil, was later discovered in some forms of nucleic acid.

DNA and RNA

The famous German scientist Emil Fischer had previously worked on the chemistry of purines and in 1901 had synthesised another chemical which was to be found in nucleic acid. This was a five-carbon sugar called ribose which was identified as a component of nucleic acid in 1910 by Aaron Theodor Levene, a Russian-born biochemist working in America. Later Levene discovered that not all nucleic acids contain ribose sugar – some have a sugar called deoxyribose instead. The pieces of the jigsaw puzzle of nucleic acid structure were coming together. Two types of nucleic acid were identified, **deoxyribonucleic acid** or **DNA** and **ribonucleic acid** or **RNA**. The purines, pyrimidines and five-carbon sugars in these nucleic acids accounted for most of Miescher's original findings of carbon, hydrogen, oxygen and nitrogen. However, the presence of phosphorus was still not explained.

Levene suggested the presence of a phosphate group, and postulated that the sugar acted like a bridge between the phosphate and the purine or pyrimidine:

Phosphate—sugar—base (purine or pyrimidine)

Levene called this arrangement of chemicals a **nucleotide** and suggested how various nucleotides might join together. The key, he said, was the phosphate group, which could link one base—sugar group to the next. In RNA the base could be adenine, guanine, cytosine or uracil. In DNA the base could be adenine, guanine, cytosine or thymine. Full confirmation of Levene's work did not come about until 1957, when the British chemist Alexander Robertus Todd was awarded the Nobel prize for chemistry for synthesising structures which obeyed Levene's formulae exactly. He found that they were identical to material obtained from biological nucleic acid. A combination of Miescher's collection of pus and Levene's intuition had led to the discovery of DNA.

The link between DNA and the cell nucleus had been firmly established many years before. German scientists of the 1880s had developed many cell-staining techniques using newly discovered dyes, which were still used in the twentieth century. The German biochemist Robert Feulgen had shown in 1924 that DNA was confined to chromosomes and not found in the rest of the nucleus. Feulgen discovered that after warming DNA with strong acid he could turn it a brilliant crimson with the dye fuchsin. When he stained sections of tissues in this way, only the chromosomes turned crimson, showing that they alone consisted of DNA. Soon it was found that all plant and animal cells had DNA in their chromosomes. Was DNA the actual material of the gene, and therefore the key to heredity? Many ingenious minds set about the task of answering this question using state-of-the-art high-technology tools of the day. In Sweden in the late 1930s, Torbjörn Caspersson used photoelectric cells and amplifiers of great sensitivity to plot the degree to which different cell components absorbed ultraviolet light. He demonstrated that nucleic acids absorb this wavelength of light very strongly, and he was able to show that nucleic acids increase in quantity wherever cells are dividing rapidly; that is, during growth. This suggested that nucleic acids have an important function in the synthesis of new cellular material.

By 1950 researchers had devised methods of estimating the quantity of DNA in a cell nucleus. Many tissues were analysed and the significant discovery was that all types of body cells in any particular species of animal had the same characteristic quantity of DNA. Thus the frog has 15 hundred-millionths of a milligram per body cell nucleus; the trout has 2 hundred-millionths of a milligram per nucleus and so on down a list of many types. In their sex cells, however, just half those amounts of DNA are present, which was significant because the sex cells formed by meiosis (see page 14) carry half as many chromosomes as do the body cells. Yet more of the jigsaw was being put in place, all indicating that DNA was closely associated with the hereditary factors in the chromosomes, and it became generally established that DNA was the chemical basis of the gene.

The behaviour of DNA

A mystery transforming agent

In 1928 the British public health officer Frederick Griffith observed some strange behaviour among bacteria. He was studying pneumonia, a widespread disease responsible for many deaths at the time. There are various types of closely related bacteria, only some of which cause pneumonia, and one way of telling them apart is by their appearance when they are grown in cultures. The bacteria form mound-shaped colonies with either a rough surface or a shiny smooth surface. The rough (R) type is harmless but the smooth (S) type can cause the disease and is therefore called virulent.

Griffith found that by injecting the S type into mice and passing blood from one mouse to another he could eventually obtain an R type. How could this change from a virulent to a non-virulent type take place? The difference between the virulent and non-virulent types is that the virulent S type has a coat of carbohydrate, whereas the R type has not. Somehow the S type had mislaid its coat while on its journey from one mouse to another. Normally both types are very stable and can be grown over many generations without change. However, about one S bacterium in every ten million will mutate and give rise to R-type colonies.

Griffith heated the virulent S-type bacteria to 60 °C and injected them into the mice. They proved to be harmless. He then injected the mice with a mixture of heat-killed virulent S bacteria and live non-virulent R bacteria. To his amazement he found that the mice sometimes died of the bacterial infection. On examining the blood of the dead mice, he found live virulent

S bacteria. When isolated and grown in cultures, these live bacteria turned out to be identical with the virulent heat-killed bacteria that had been originally injected. He deduced that some factor must have passed from the heat-killed virulent S bacteria to the living non-virulent R bacteria, causing them to change their nature. After several attempts by others to confirm Griffith's results, it became clear that a transforming agent had passed between the bacteria. Although this chemical was normally contained in the bacterial cell, it could act independently of it. In fact the transforming agent was the genetic material responsible for handing down characteristics from one generation to the next. The next step was to isolate this chemical and unravel its chemistry.

Along came a shy sensitive American called Oswald Theodore Avery. He was qualified as a medical doctor but he had decided to devote himself to research rather than practise medicine. He was devoted to his research, to the extent that he chose to live directly opposite the Rockefeller Institute so that he was as close as possible to his laboratory. He remained a bachelor, never took holidays and even stopped answering correspondence because it 'wasted precious hours stolen from research'!

Avery had studied the pneumonia-causing bacterium from as early as 1913. Griffith's results fired his imagination and in the early 1940s he teamed up with Colin Munro Macleod and Maclyn McCarty to track down the so-called transforming agent. First they selectively attacked the known chemical components of the bacterial cells with enzymes, one by one. In this way they hoped to arrive at the transforming agent by a process of elimination. They attacked the carbohydrate coat of the S-type bacteria using enzymes. Even when the coat was broken down in this way the remaining cell was still a powerful transforming agent, proving that the agent was not the carbohydrate coat. They then attacked the proteins in the bacteria with protein-splitting enzymes. Again the transforming properties of the cells were unaffected. Of the known materials left, the only significant ones were the nucleic acids DNA and RNA.

DNA – the chemical that carries the genes

Now came the decisive experiment. The research team took a solution of the transforming agent and added a purified enzyme which was known to break down DNA specifically. After such treatment the transforming agent failed to be effective. They had proved beyond doubt that the transforming agent was DNA. In 1944 they published their results:

> The active transforming material … contains no demonstrable protein, unbound lipid, or serologically reactive polysaccharide, and consists principally, if not solely, of a highly polymerised viscous form of deoxyribonucleic acid.

Avery and his team had isolated the genetic material of cells. Here for the first time a chemical basis had been found to explain the actual mechanism of inheritance so meticulously studied by Mendel. The discovery of Avery's team astounded other biologists working in the same field of research. Previously it was thought that proteins were the chemicals of inheritance because they were known to be part of the fabric of chromosomes. Also proteins were recognised as being complex enough to carry the vast number of instructions necessary to account for the inheritance of characteristics. The doubters carried out numerous replica experiments to confirm the findings. By 1955 some thirty different examples of similar genetic transformations had been induced in other types of bacteria. The doubters now became believers and were at last convinced of the hereditary role of DNA in bacteria. However, was the same true in other organisms?

Evidence that DNA works in a similar way in other organisms was given by a type of bacteriophage. A tadpole-shaped T phage virus infects the bacterium *Escherichia coli*, as shown in Figure 3.2. When a suspension of the T phage is added to a culture of *E. coli*, clear areas appear in the culture where the viruses have infected the bacteria.

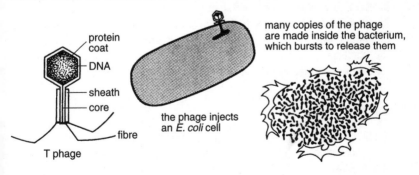

Figure 3.2 The infection of a bacterium by a bacteriophage

The viruses attach to the surface of the bacterium by their tails. About 20 minutes later the bacterium bursts and releases hundreds of new viruses. The genetic material of the infecting virus passes into the bacterium and once there causes many copies of the virus to be made. During this process the bacterium is destroyed. Attempts were made to identify the genetic material inside the virus by removing the contents of viruses and analysing them chemically. Roger Moss Herriott of Johns Hopkins University succeeded at this, and discovered that the contents were mostly DNA and a little protein. This experiment suggested that the virus genes were made of DNA or protein.

An ingenious experiment to finally prove that genes are made of DNA was performed in 1952. The researchers were Alfred Day Hershey and Martha Chase, working at the Cold Spring Harbour genetics laboratory of the Carnegie Institution. They used radioactive isotopes as tracers to follow the movement of both the virus protein and its DNA. Their work depended on the fact that DNA contains phosphorus but not sulphur, and that the virus protein contains sulphur but not phosphorus. The virus DNA can therefore be labelled with a radioactive isotope of phosphorus and the virus protein with a radioactive isotope of sulphur.

To label the viruses in this way, Hershey and Chase started by growing *E. coli* in a medium containing either radioactive sulphur or radioactive phosphorus. The bacteria took up the radioactive elements from the media and incorporated them into their cells. The labelled bacteria were then infected with viruses and the new viruses formed as a result were found to be labelled with radioactive phosphorus or radioactive sulphur. These labelled viruses were now used to infect unlabelled bacteria. When sulphur-labelled viruses were used, the bacteria did not become radioactively labelled. However, following infection by the phosphorus-labelled viruses the bacteria were found to be radioactively labelled. This proved that phosphorus-containing DNA had passed into the bacteria.

The structure of DNA

The importance of DNA now being recognised, the scene was set to determine the detailed structure of the DNA molecule. Erwin Chargaff of Colombia University carefully analysed the relative amounts of the four bases known to occur in DNA. He examined the DNA from a range of species and by 1947 he had found that the two purines and the two pyrimidines were present in widely varying amounts. Hence there could not be any simple sequence of four nucleotides repeated over and over. He also found that the proportions of purines and pyrimidines differed a great deal from one organism to another; each organism had its own specific proportions. Most important, he discovered that the ratio of adenine to thymine was always one to one, and so was the ratio of cytosine to guanine. Later the significance of this became apparent when explaining the constant width of the DNA molecule, but at the time little was known about the three-dimensional nature of DNA.

X-ray diffraction reveals the helix

Physicists joined the hunt for the molecular shape of DNA using the technique of X-ray diffraction. If a beam of X-rays shines through a crystal in which the molecules are regularly arranged, it is split up or diffracted in a way that depends on the arrangement of the atoms in the molecules. By passing the diffracted X-rays to a photographic plate, a complex pattern of light and dark results. The light areas indicate where the X-rays have struck the developed plate. From this pattern scientists can work out how the beam was diffracted, and consequently find how the atoms are arranged. With the help of a computer and some mathematical analysis of the data, the shape of the crystal can be worked out.

The snooker analogy

An analogy of X-ray diffraction is a blind snooker player shooting balls at random across a snooker table. By measuring the angle at which the balls come back and counting those that never return, it is possible to work out the position of the pockets.

The technique of X-ray diffraction was used as long ago as the early 1930s when William Astbury at the University of Leeds made diffraction pictures of proteins. By 1950 Linus Pauling and Robert Brainard Corey had deciphered the three-dimensional nature of some proteins as being helical (spiral). This pioneering X-ray diffraction work to find helical protein molecules was to be vitally important in the eventual discovery of the structure of DNA.

At about the time that Pauling and Corey were making their discoveries, Maurice Hugh Frederick Wilkins and Rosalind Franklin began much more detailed crystallographic studies of DNA at King's College, London.

Wilkins was a New Zealander and during the Second World War was a physicist working on nuclear weapons. In post-war years he turned his interest to biology and became one of the first of a new kind of scientist – a biophysicist, exploring the physics of living matter. Rosalind Franklin was an intensely serious and highly skilled crystallographer who contributed a great deal to the initial ideas of DNA structure but who died before the completion of the puzzle.

Watson and Crick – making models

In 1953 two scientists produced an hypothesis which resulted in one of the most famous discoveries of all time. The British researcher Francis Harry Compton Crick collaborated with the young American James Dewey Watson in a partnership which proved to be perhaps the most rewarding example of Anglo-American co-operation in the history of science.

Crick had spent the war years designing mines for the Royal Navy and had then changed direction and become fascinated by molecular biology. He worked with X-ray diffraction techniques on large protein molecules in America, but then moved to Cambridge where he met Watson who had a research fellowship in biology. A brilliant unpredictable young man, Watson had been a child prodigy, studying at a college in America at the age of fifteen. This dynamic duo of physicist and biologist set about building many models of DNA until they made one which fitted the known facts. They meticulously constructed exact scale models with precise lengths and angles to match the lengths and angles of the chemical bonds in the molecule.

Their first problem was to decide how many chains made up the molecule – earlier suggestions had been two or three. They drew on the research of Wilkins who had found that the width of DNA was a constant 2 nanometres.

They applied some clever logic to this knowledge. They assumed that ten nucleotides fitted into every 3.4-nanometre length of molecule, a fact which had been indicated by the X-ray diffraction data. Then they worked out the density of the molecule from the masses of the atoms involved. They started by assuming there was one chain of nucleotides, and this gave a result that was half the known density. Hence they concluded that the molecule must have twice as many nucleotides in its length, so that it must consist of two chains. They now used Chargaff's discovery of 1947; that in any DNA molecule there were equal numbers of adenine and

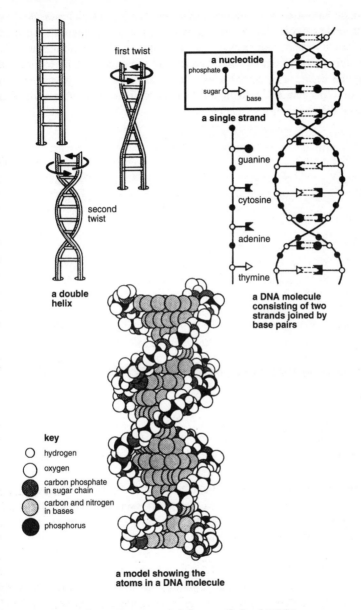

first twist

a nucleotide

phosphate ●

sugar ○——▷
                  base

a single strand

guanine

cytosine

adenine

thymine

second
twist

**a double
helix**

**a DNA molecule
consisting of two
strands joined by
base pairs**

key

○ hydrogen

○ oxygen

● carbon phosphate
  in sugar chain

○ carbon and nitrogen
  in bases

● phosphorus

**a model showing the
atoms in a DNA molecule**

Figure 3.3 The Watson–Crick model of DNA

thymine bases and equal numbers of guanine and cytosine bases. So they built models of two nucleotide chains to see how these might fit together. Crick and Watson suggested that the bases paired with each other something like dominoes; one number had to be matched by its partner. Adenine paired with thymine and guanine paired with cytosine. Crick wrote: 'This pairing is likely to be so fundamental for biology that I cannot help wondering whether some day an enthusiastic scientist will christen his newborn twins Adenine and Thymine.'

With a brilliant flash of insight, the two partners in discovery used the knowledge of the constant width of the DNA molecule to show how the two bases could point inwards to form a pair. They decided that the base pairs formed the rungs of a spiral ladder, as shown in Figure 3.3.

Amazingly the whole model was worked out within six weeks! It is said that during the work Watson continued his habits of having a leisurely cafe breakfast each morning and a game of tennis each afternoon. The term **double helix** is used to describe the final shape of two chains of nucleotides twined around each other like a rope ladder. The work was published in 1953 in one of the most famous papers ever to appear in the scientific journal *Nature*. There was immediate worldwide appreciation of the sheer elegance and simplicity of Crick and Watson's work. Wilkins, Crick and Watson received the Nobel prize for medicine and physiology in 1962. Crick built a symbolic copper helix outside his house in the centre of Cambridge. What a contrast to the cool reception of Mendel's famous paper in 1865 where it all began!

How genes work

Genes regulate the manufacture of proteins by cells. At any given time there are thousands of chemical reactions going on in our bodies, and they are all controlled by enzymes. Enzymes are proteins, and it is by determining which enzymes are produced that genes exert their control over our chemistry. So in order to make sense of the way in which genes work, we must look at the nature of proteins and enzymes.

An introduction to proteins

Proteins are complex molecules found in a whole range of structures of our bodies. Hair, horn, muscle, silk and tendon have structures made up of protein. Less visible forms of protein regulate our biochemistry and therefore determine whether we live or die. Haemoglobin carries oxygen in blood; hormones carry chemical messages around the body; and most important of all, biological catalysts called enzymes are essential for the series of controlled chemical reactions that we call life.

The name **protein** was coined in 1838 by Gerardus Johannes Mulder. He realised that protein was the most important organic chemical yet discovered in the body, and gave it a name that meant 'primary' or 'holding first place'. Besides being responsible for the basic chemistry of life processes, proteins also give us individuality – every organism forms its own kinds of protein. As with fingerprints, no two individuals have identical sets of proteins, since their synthesis is under the control of unique combinations of genes.

The structure of proteins was first studied in detail in the 1870s by one of the most outstanding chemists of all time, the German Emil Fischer. As far back as 1819 a French chemist, M H Braconnet, had isolated **amino acids**, the building bricks of proteins. He boiled the protein gelatin with acid and obtained glycine, the simplest amino acid. Many other amino acids were isolated during the nineteenth century. Fischer showed how these units linked together in chains joined by bonds called peptide linkages. Chains of amino acids joined in this way are called **polypeptides** – many (poly) peptides. It is known today that there are twenty amino acids found in proteins and the number of different ways that they can be combined is astronomical. A medium-sized protein molecule is made up of about 5000 atoms in total. These giant molecules were so difficult to analyse that progress was not really made until the 1940s when Charles Chibnall and his colleagues at Cambridge University attempted to work out their structure. Frederick Sanger spent ten years discovering the amino acid sequence in insulin, a relatively small protein with only 51 amino acids. He was awarded a Nobel prize for this work. The sequences of many more proteins are now known.

Enzymes – special proteins

Our bodies are chemical factories that carry out thousands of chemical reactions with a precision, speed and efficiency that could never be rivalled by chemists in laboratories. The laboratory chemist will take months to synthesise relatively simple organic compounds. By contrast, a single bacterium can synthesise all the chemicals that it needs to make a copy of itself in twenty minutes!

The chemical reactions taking place in all our cells are carried out in finely tuned sequences. Each individual reaction step in any sequence is controlled by a specific chemical **catalyst**, a substance which helps the reaction to happen without being changed itself by the reaction. The catalysts made by living things which speed up the body's chemical reactions are **enzymes**.

Many students' early perceptions of the word enzyme are linked with digestion, because the first introduction to enzymes is usually in the context of the digestive system. The history of the study of enzymes also began with digestion. However, the vast majority of enzymes have nothing whatsoever to do with digestion – the digestive enzymes make up a relatively small group.

One of the earliest examples of enzyme action to be discovered was that of pepsin, the protein-digesting enzyme produced in the stomach. The remarkable Spallanzani, famous for putting silk pants on toads (see page 11), investigated digestion in the eighteenth century. He fed hawks with meat enclosed in small wire capsules. He later made the birds regurgitate the capsules and found that they no longer had meat in them. He concluded that, since the meat could not have been ground up inside the birds, it had been liquefied by something in the stomach juices. In 1836 that something was identified as pepsin.

Even before this, in 1830, Anselme Payen had discovered that malt extract contained a substance that could break down starch to sugar. He called it diastase, and the following year a similar substance with similar properties was found in saliva. This was ptyalin, now called amylase.

The newly discovered chemicals produced marked effects on chemical reactions even when they were present in minute amounts. For example rennin, which is an enzyme produced in the stomachs of all baby mammals, can clot ten million times its own weight of milk in five minutes. Even more spectacular is the enzyme catalase, which can break down over two million molecules of hydrogen peroxide per minute. This action is vital to us because our cells produce damaging hydrogen peroxide as a waste product of metabolism. Catalase quickly breaks it down to harmless oxygen and water. The action of catalase on hydrogen peroxide has been used in surgical dressings for use with military casualties. Oxygen is produced in the dressings, which kills certain disease-causing microbes in wounds.

The idea of catalyst action came from the famous nineteenth century Swedish chemist Jöns Jakon Berzelius who pointed out that many chemical reactions happened quickly enough to be noticeable only if a relatively small amount of some other chemical were present, which was not itself affected chemically by the reaction. He used the example of hydrogen and oxygen reacting together to produce water in the presence of the metal platinum. By 1835 the idea of catalyst action was extended to chemicals made by living organisms, leading to the concept of an enzyme. With astounding insight, Berzelius anticipated that all substances in living organisms were produced by the action of enzymes.

By definition, an enzyme is a protein made by living things that can speed up a chemical reaction. In 1907 the German Eduard Buchner was awarded a Nobel prize for demonstrating that enzymes extracted from yeast could carry out fermentation. Previously the fermentation reaction had been carried out only by whole yeast cells.

The ability of enzymes to catalyse reactions seemed to be magical, but the myth was destroyed in the 1920s when James Batchellor Sumner at Cornell University attempted to isolate urease as a pure enzyme from jack beans. An enzyme is commonly named after the substrate it acts on with the ending -ase. Urease acts on urea, a waste substance found in urine, and breaks it down to ammonia and carbon dioxide. Sumner's extracted urease was in the form of octagonal crystals, which he found to be pure protein. He had succeeded in crystallising an enzyme for the first time. There was nothing mystical or magic about this enzyme – it was just a pure protein. However, his rivals in the scientific circles of the day would not accept his findings and several leading authorities dismissed his work. It was not until 1947 that the importance of Sumner's discovery was recognised by the award of a shared Nobel prize for chemistry, after another enzyme, pepsin, was also proved to be a protein.

All the cells making up an organism must work together for it to remain alive. The harmonious regulation of these cells is due to the orderly work of enzymes, and the mechanism for this regulation must be transmitted from one generation to the next. Hence genes must regulate the production of enzymes, and the next step towards understanding how genes work was to comprehend their relationship to enzymes.

One gene, one enzyme

Human hereditary enzyme deficiencies – circumstantial evidence

The suggestion that a gene might be responsible for controlling the production of a specific enzyme was made by Archibald Garrod in 1908. A rare inherited disorder concerning the metabolism of urine had been recognised as long ago as the seventeenth century, in which an affected person's urine turns black when exposed to air. In 1859 a German scientist, C Bodeker, found that the darkening was caused by a chemical called alcapton. The disorder became known as alcaptonuria, meaning alcapton in the urine. It is inherited as a recessive mutant gene. Garrod suggested that the alcapton was present because an enzyme was missing, and the enzyme was missing because of the defective gene.

Alcaptonuria is not life threatening or particularly distressing, but other enzyme problems caused by single defective genes have much more terrible consequences. One such disorder is phenylketonuria (PKU), in which the defective gene prevents the formation of an enzyme that changes the amino acid phenylalanine to another amino acid called tyrosine. Phenylalanine is essential in our diet, but if it is not converted to tyrosine it builds up in the blood and can damage the brain.

In the 1930s scientists explained PKU's inheritance as a single recessive gene, so the disease only affects homozygous recessives. Affected newborn children are given a carefully prepared diet containing just enough phenylalanine for their needs.

A third example of an inherited error of metabolism in which a single gene defect means an enzyme is missing is galactosaemia. This is a rare condition in which sufferers cannot use galactose sugar, produced in the body when milk sugar, lactose, is digested. As a result there is a build-up of galactose in an affected baby's cells which can lead to brain damage. This inability to metabolise galactose is because of the lack of a single enzyme. Normal milk must be excluded from such an infant's diet, assuming the problem can be diagnosed early enough to prevent damage (see gene profiling on page 135).

Inducing mutations in a mould

More exact knowledge about the link between genes and enzymes came from research carried out by the Americans George Wells Beadle and

Edward Lawrie Tatum. They carried out classic investigations in the 1930s with a pink mould – a fungus called *Neurospora* which does not have a common name but is often found growing on damp stale bread. Beadle gained his doctorate in 1931 as a geneticist, having been fascinated by Thomas Hunt Morgan's investigations into the genetics of fruit flies. In the autumn of 1937 Beadle was working at Stanford University, where he met Tatum. They searched for a gene mutation that would influence the chemical reactions taking place in living things. They chose *Neurospora* as their research material because its diet consists of sugar, mineral salts and biotin, one of the vitamin B group. With this simple diet *Neurospora* can synthesise all of the twenty amino acids and all the vitamins it needs other than biotin.

Because *Neurospora* can synthesise so many substances it followed that the fungus must have genes to govern the production of all the enzymes involved. Beadle and Tatum predicted that by causing some of the mould's genes to mutate they should be able to produce moulds with errors of metabolism that could be identified chemically. Beadle's analogy said:

> If cars are observed as they emerge from an assembly line, it is not possible to determine what each worker did as part of the cars' manufacture. However, if one could replace able workers one by one with defective ones and then observe the result in the products, it would be possible to determine that one worker put on the radiator, another added the fuel pump, and so on.

At the time of Beadle and Tatum's work, a whole arsenal of mutation-causing agents was available. Muller had already succeeded in producing hereditable mutations using X-rays. Phosgene (mustard gas) and ultra-violet light were also know mutagens, so there was no shortage of methods for producing mutations in laboratories.

Like most other fungi, *Neurospora* reproduces both sexually and without sex (asexually). In asexual reproduction there is no mixing of genes or other genetic change, so Beadle and Tatum had a potentially unlimited supply of pure-breeding strains of the mould. During sexual reproduction, however, there is fusion of the gametes during fertilisation, with consequent mixing of genes. Each fused nucleus then divides to produce eight spores (the equivalent of seeds of flowering plants) neatly arranged in a line in club-shaped spore sacs. If a single gene is mutated in the parent strain, four of these spores will carry the mutated gene while the other four spores will carry the normal gene. A scientist skilled in micro-dissection can separate the spores from the spore sac and therefore plant out separate

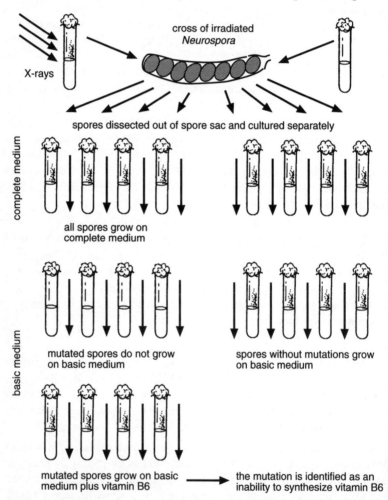

Figure 3.4 Beadle and Tatum's procedure, which proved that one gene is responsible for one enzyme

pure cultures. By controlling the nutrients available to the cultures, the dietary requirements of each strain of mould can be determined.

Beadle and Tatum used the following procedure illustrated in Figure 3.4.

- They subjected asexually produced pure-bred strains of the mould to X-rays and allowed these to reproduce sexually so that they produced spore sacs.
- They planted each of the eight spores separately on a complete medium – a mixture of nutrients containing everything the mould would need for growth, including substances which it would normally make for itself. At this point any deficiency that the mould might have in making any substances could not show up.
- Samples of the resulting growths of mould were then placed on the basic nutrient medium needed by a normal mould.

Beadle and Tatum reasoned that if a mutation had taken place, four of the eight samples would not grow on the basic medium because of their genetic defect. They began a painstaking programme to check a range of irradiated samples. Once they had found a defect in a particular sample, they tested sub-samples on other basic nutrient media each containing one added nutrient that the mould would normally make. In this way they would eventually find one nutrient medium that would allow the mould to grow, and the added substance it contained would be the one the mutant mould could no longer make. Following meticulous sampling they identified their first mutant in sample number 299. This needed vitamin B6 in addition to the basic diet. A second mutant which needed vitamin B1 was found in sample 1085. Fortunately dozens more were found quite soon afterwards. Like the early founders of genetics such as Mendel and Morgan, their work needed unlimited patience and care.

Having found mutant moulds with precise metabolic defects, Beadle and Tatum needed to know whether each defect was a result of a mutation of only one gene or of more than one. The logic Beadle used was:

> It was a simple matter to determine whether our newly induced deficiencies were the result of mutations in single genes. If they were, crosses with the original should yield four mutant and four non-mutant spores in each spore sac. They did.

As a result of this work Beadle and Tatum became famous for the one gene, one enzyme concept and received a share of the 1958 Nobel prize for medicine and physiology.

Sickle-cell anaemia

What is the relevance of this discovery to ourselves, or to any organisms other than moulds? This close examination of the control exercised by a single gene has made a world of difference to sufferers of the life-threatening disorder sickle-cell anaemia. This is a blood defect which exists among races native to northern Africa and the Mediterranean countries of southern Europe. It is so severe that affected people usually die young.

In sickle-cell anaemia the haemoglobin in the red blood cells is defective, causing the cells to be abnormally shaped. Normally human red blood cells are disc-shaped with bowl-like depressions on each surface. People with sickle-cell anaemia have crescent-shaped (sickle-shaped) red blood cells which cannot carry as much oxygen as normal. The pattern of inheritance of the condition was not worked out until 1949, when James Van Gundia Neel at the University of Michigan showed that severe anaemia occurs when the person has a pair of mutant alleles (is homozygous). If only one of the pair of alleles is defective (in a heterozygote), the result is called sickle-cell trait which is much less severe and makes little difference to a person's normal activities. Indeed people who are heterozygous for this gene are more resistant than normal to malaria, which is caused by a parasite that spends part of its life cycle inside red blood cells. The inheritance of sickle-cell anaemia and sickle-cell trait is shown in Figure 3.5, where **Hb** is the allele for normal haemoglobin and **HbS** the sickle-cell allele.

The analysis of normal and sickle-cell haemoglobin was carried out by Linus Pauling using the novel technique of electrophoresis. Haemoglobin was placed on paper soaked in a conducting solution and exposed to an electric field. Normal haemoglobin moves to the positive pole, sickle-cell haemoglobin moves to the negative pole and haemoglobin from a person with sickle-cell trait separates, some moving to the negative pole and the rest moving to the positive pole. The haemoglobin molecule is made of a protein part and an iron-containing part, and Pauling showed that the difference between the two haemoglobins was in the protein part of the molecule. Detailed analysis showed that out of a chain of some 300 amino acids making up the haemoglobin molecule, only one amino acid was different. This apparently minor molecular change is enough to make the difference between life and death, and it is caused by a defect in a single gene. Here was undoubted evidence for the chemical nature of the control of genes over the development and functioning of the body.

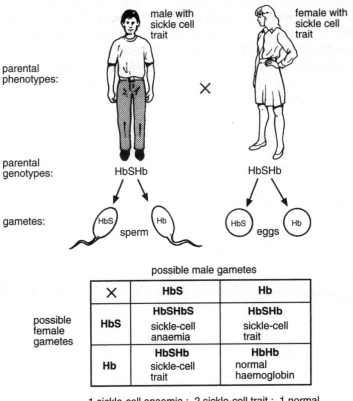

1 sickle-cell anaemia : 2 sickle-cell trait : 1 normal

Figure 3.5 The inheritance of sickle-cell anaemia

How DNA copies itself

One important implication of Watson and Crick's discovery of the double helical structure of DNA (see page 92) was that it suggested a mechanism for self-duplication or replication. A quotation from Watson and Crick clearly shows that they had recognised this potential copying mechanism:

It has not escaped our notice that the specific pairing we have postulated immediately suggests a possible copying mechanism for the genetic material. The pairing rule of bases in the DNA molecule ensures that one nucleotide sequence of one chain in the double helix must have a fixed relation to the other chain. Therefore, if the sequence in one part of the fixed chain is ATAGGC, the sequence along the second complementary chain must be TATCCG, where A = adenine; T = thymine; G = guanine; and C = cytosine. If the two nucleotide strands separate, each will serve as a template to govern the structure of the other.

With this in mind, Crick suggested that the two spirals in the DNA molecule might begin to unwind and separate in the presence of a supply of new nucleotides. The new nucleotides would attach themselves by their bases to the corresponding partners on the template, as shown in Figure 3.6. Each newly paired nucleotide would remain attached, another would fit alongside and gradually a new nucleotide chain would grow until a complete complement to the first was formed. Eventually two DNA molecules would exist where only one existed before.

One difficulty in accepting this idea was the need for the whole DNA molecule to unwind into its halves before duplication could begin. As Crick pointed out, a single chromosome must be equivalent to around ten million turns of the spiral. The idea of all this unwinding before duplication begins is hard to believe. Crick predicted that the synthesis of two new chains might begin as soon as the original ones had started to unwind.

Direct confirmation that DNA is a self-replicating molecule was established in 1956 by Arthur Kornberg at Washington University in St Louis. Kornberg succeeded in synthesising DNA and proved conclusively that, with the help of an enzyme, DNA could copy itself. He was awarded the Nobel prize for medicine and physiology in 1959.

Several scientists were anxious to confirm the Watson–Crick hypothesis and some ingenious experiments were carried out in the late 1950s. Among these was the highly elegant investigation by Matthew Stanley Meselson and Franklin William Stahl at the California Institute of Technology, published in 1958. They used the bacterium *Escherichia coli*

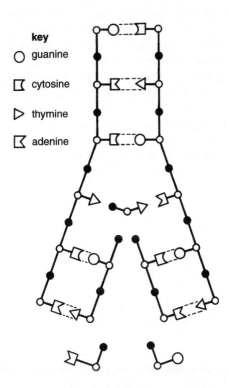

Figure 3.6 How the DNA molecule replicates – the two strands of the DNA molecule begin to separate and new nucleotides fall into place by base pairing

and grew it in a culture solution containing glucose, mineral salts without nitrogen and ammonium chloride in which all the nitrogen was 'heavy nitrogen', isotope N-15, instead of the normal nitrogen, N-14. After allowing the bacteria to divide fourteen times, Meselson and Stahl had obtained bacteria with heavy nitrogen in their DNA. They then changed the medium by adding an excess of ordinary ammonium chloride containing N-14, so that in future growth the DNA formed by the bacteria would have ordinary N-14 in it.

If the Watson–Crick model were correct, they reasoned, all bacteria formed in the first cell division after the addition of ordinary ammonium chloride would have DNA molecules, half containing the original heavy nitrogen and half containing the new ordinary nitrogen. At the second cell division, half of the offspring would contain some heavy nitrogen while half would have only ordinary nitrogen. To find out whether this had indeed happened Meselson and Stahl had to check the weight of the DNA at successive divisions of the bacteria to see if the DNA molecules contained heavy nitrogen. An astounding degree of accuracy was called for and a technique involving ultracentrifugation was used. If a solution is rotated in a tube at extremely high speeds, heavier molecules are thrown towards the bottom of the tube. A density gradient is formed, with the solution being most dense at the bottom and least dense at the top. This was the technique Meselson and Stahl used to discover the density of DNA extracted from the bacteria before and after the addition of ordinary ammonium chloride to the culture medium.

Their first sample came from bacteria just before the addition of ordinary ammonium chloride to the culture medium. It was therefore composed of virtually nothing but heavy-nitrogen DNA. The next sample came from bacteria which had divided once after the addition of ordinary ammonium chloride. This DNA was less dense. The third sample came after two bacterial divisions and contained two densities of DNA – one the same as the second sample and one of ordinary density. These two densities of DNA were present in equal amounts. Finally, after four divisions, there was a trace of the intermediate-density DNA, but most of the DNA was of ordinary density. They interpreted the intermediate-density DNA as being made of one nucleotide strand containing heavy nitrogen and one nucleotide strand containing ordinary nitrogen. The method of copying had conserved just part of the original DNA molecules. The concept became known as **semiconservative replication**.

Once it had been established that gene action worked at the molecular level, thoughts were turned to the questions of how DNA controls parts of the cell outside the nucleus, and how it regulates the formation of enzymes and other proteins when they are needed.

How DNA controls protein synthesis

The role of RNA

It was not until the end of the 1930s that scientists realised that ribonucleic acid or RNA was present in both animal and plant cells. Before this it was thought that DNA was a component of animal cells and that RNA was to be found only in plant cells. The Belgian Jean Brachet and the Swede Torbjörn Caspersson were the first to show the universal presence of RNA in living cells in the 1940s. They also found that most of the RNA occurs outside the nucleus, although some is contained in special parts of the nucleus called nucleoli. This discovery guided them into thinking that RNA might form some sort of link between the nucleus and the rest of the cell. Caspersson and Brachet continued their research and discovered that cells that specialise in making lots of protein are those with most RNA. For example, cells from the glands of the silk moth that make silk, which is pure protein, have an abundance of RNA. It could be possible that the synthesis of protein depended on the presence of RNA. By 1943, ultracentrifuges had been designed to spin fast enough to produce 18 000 times the force of gravity. One of these was used by Albert Claude of the Rockefeller Institute to separate the smallest parts of cells. He managed to produce a sediment of the smallest cell components, the ribosomes, which are only visible under an electron microscope. Fine chemical analysis showed that the ribosomes were a mixture of RNA and protein, and the RNA in these particles accounted for most of that found in the whole cell.

In 1950 work began to identify the site in the cell where protein was made. After exposing cells to amino acids labelled with radioactive carbon followed by ultracentrifugation, researchers found that ribosomes had more than twice as much radioactivity per gram of protein as the other cell components. This indicated that the ribosomes were sites for the assembly of protein from their amino-acid building blocks. In the mid-1950s it was shown that an enzyme which broke down RNA prevented protein synthesis in the cell, but when more RNA was added protein synthesis started again. All this research demonstrated conclusively that RNA at the site of the ribosomes was essential for making proteins. The questions that remained were how did RNA operate, and what was the role of the rest of the RNA in the cell?

Transfer RNA

Confusion gave way to clarity in 1957. By then over twenty different kinds of soluble RNA had been identified, each linked to one kind of amino acid. It was shown that the soluble RNA, or **transfer RNA** as it was called, brought amino acids to the ribosome and lined them up in the correct sequence to form a particular protein.

From physical and chemical studies it is now known that transfer RNA contains only a single nucleotide chain of about 80 nucleotides. The complete nucleotide sequence of a transfer RNA molecule was first elucidated in 1965 by a team of scientists at Cornell University under the direction of Robert W Holley. Several modifications to their model have since been made, but the essential role of transfer RNA molecules in protein synthesis remains that they combine with specific amino acids to bring them in a line at a ribosome.

Messenger RNA

Although the discovery of transfer RNA was very important, it did not solve the riddle of how DNA controls the formation of proteins. Scientists reasoned that some complementary pattern or template must exist to ensure that the transfer RNA molecules brought their amino acids into line in the correct sequence for making a particular protein. In the early 1960s a brilliant suggestion welded together all the previous observations in this field and made a framework for an explanation. The stroke of genius came from François Jacob and Jacques Monod at the Pasteur Institute, Paris. They boldly proposed that a newly discovered RNA molecule that had a DNA-like structure had the function of carrying genetic instructions from the nucleus to the ribosomes. They named this special type of RNA **messenger RNA**. The hypothesis offered a working model for the vital transmission of information from DNA to the protein assembly area. More researchers directed their efforts towards this idea and soon a picture emerged in which messenger RNA was being continually sent out from the genes to the ribosomes, where it was used and then replaced by fresh messenger RNA.

The genetic message was thought to be transferred from DNA to messenger RNA by a base-pairing method. The first step must be for the two spirals of the DNA molecule to separate, exposing the bases. A complementary strand of messenger RNA could then be built alongside one of the DNA strands in a process called **transcription**. In 1963 messenger RNA was proved to be a single stranded molecule transcribed from only one strand in the DNA molecule. An enzyme called **RNA polymerase** splits the two spirals of DNA, like someone undoing a zip fastener, thus allowing the RNA to build up along the exposed bases in one chain. In the base pairing in RNA a new base uracil substitutes for thymine. In this way the genetic message embodied in one DNA strand is used as a code for the synthesis of messenger RNA, which itself codes for the amino acids to be brought to the ribosomes in order to make a protein.

The genetic code

There is a genetic alphabet which is used to translate the linear sequence of bases in DNA into the sequence of amino acids in proteins. The alphabet is A for adenine, U for uracil, C for cytosine, and G for guanine. Unbelievable as it may seem, an alphabet of only four bases is enough to code for the thousands of different proteins that exist in living organisms. You will recall that although proteins are enormous and complex molecules, they are built of twenty different types of amino acids (see page 93). There is a simple mathematical explanation. If each base codes for one amino acid, only four amino acids can be specified. A two-base combination would provide four times as many possible combinations ($4 \times 4 = 16$), but this is still not enough. By 1952, which was quite early in the search for the genetic code, the biochemist Alexander Latham Dounce predicted a three-base combination to code for each amino acid. The prediction was confirmed by George Gamow, a physicist, two years later. It was reasoned that a combination of any three of the four bases would provide 64 possible triplets ($4 \times 4 \times 4 = 64$). However, there are only twenty amino acids, so more than one triplet of bases could code for the same amino acid. Each triplet of bases is called a **codon**. The genetic code is shown in Table 3.1.

Sixty-one of the sixty-four codons code for a specific amino acid. The three remaining codons are UAA, UAG and UGA. These act like full stops at the end of a sentence. In addition, the codon AUG signals the beginning of a message, like a capital letter.

Coding for protein synthesis

You are now aware of the nature of DNA, messenger RNA and transfer RNA. Their structures and functions are summarised in Table 3.2. We can now follow the synthesis of a protein as determined by the genetic code.

A messenger RNA molecule is formed alongside a DNA strand in the nucleus. It leaves the nucleus and enters the cytoplasm, diffusing through the cell until it reaches a ribosome. It joins with the ribosome where transfer RNA molecules bring amino acids to it. There are as many types of transfer RNA molecules as there are amino acids, and each type of transfer RNA molecule can attach to only one type of amino acid. This specificity of the transfer RNA molecule is determined by three nucleotides called an **anticodon** found at one end of the transfer RNA molecule. For example, if the anticodon bases all are adenine, AAA, this anticodon pairs

Table 3.1 The genetic code – sixty-four triplet combinations (codons) each code for an amino acid

Second letter (base)

First letter (base)	U	C	A	G	Third letter (base)
U	UUU phenylalanine	UCU serine	UAU tyrosine	UGU cysteine	U
	UUC phenylalanine	UCC serine	UAC tyrosine	UGC cysteine	C
	UUA leucine	UCA serine	UAA stop	UGA stop	A
	UUG leucine	UCG serine	UAG stop	UGG tryptophan	G
C	CUU leucine	CCU proline	CAU histidine	CGU arginine	U
	CUC leucine	CCC proline	CAC histidine	CGC arginine	C
	CUA leucine	CCA proline	CAA glutamine	CGA arginine	A
	CUG leucine	CCG proline	CAG glutamine	CGG arginine	G
A	AUU isoleucine	ACU threonine	AAU asparagine	AGU serine	U
	AUC isoleucine	ACC threonine	AAC asparagine	AGC serine	C
	AUA isoleucine	ACA threonine	AAA lysine	AGA arginine	A
	AUG methionine (start)	ACG threonine	AAG lysine	AGG arginine	G
G	GUU valine	GCU alanine	GAU asparagine	GGU glycine	U
	GUC valine	GCC alanine	GAC asparagine	GGC glycine	C
	GUA valine	GCA alanine	GAA glutamine	GGA glycine	A
	GUG valine	GCG alanine	GAG glutamine	GGG glycine	G

Table 3.2 Summary of the structure of nucleic acids

Nucleic acid	Structure	Function
DNA (deoxyribonucleic acid)	Double twisted strand of deoxyribose (sugar), phosphate and base. Bases are adenine, cytosine, thymine and guanine.	Holds genetic information in its base sequence.
RNA (ribonucleic acid)	Strand of ribose (sugar), phosphate and base. Bases are adenine, cytosine, uracil and guanine.	Messenger RNA (mRNA) carries DNA messages in its sequence of of bases (codons) from the nucleus to the cytoplasm.
		Transfer RNA (tRNA) is an adaptor molecule that brings the correct amino acid to the mRNA molecule.

with a codon with three uracils, UUU. The only amino acid that this transfer RNA molecule can attach to is phenylalanine (see Table 3.1). This transfer RNA molecule carrying a phenylalanine amino acid meets the messenger RNA at a ribosome, and the only place it can fit is where there is a codon of three uracils, UUU. This is how both the type of amino acid in a protein and its position are dictated by the DNA in the nucleus. Chains of amino acids linked together by peptide links eventually build up to form proteins, which may be structural proteins or enzymes. An enzyme formed in this way will control one of the thousands of biochemical reactions that go on in the body at every instant, and will determine the very nature of a particular cell. The process of transcription of the code from DNA to messenger RNA and the subsequent protein synthesis is shown in Figure 3.7.

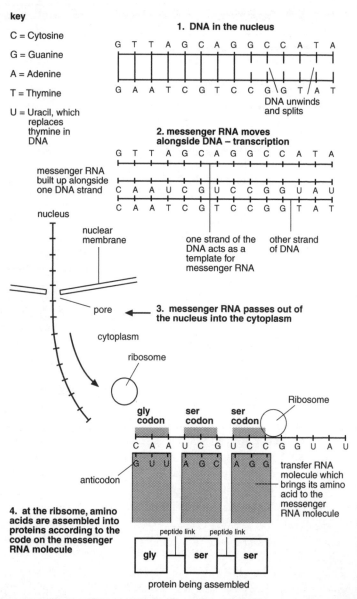

key

C = Cytosine

G = Guanine

A = Adenine

T = Thymine

U = Uracil, which replaces thymine in DNA

1. DNA in the nucleus

DNA unwinds and splits

2. messenger RNA moves alongside DNA – transcription

messenger RNA built up alongside one DNA strand

nucleus

nuclear membrane

one strand of the DNA acts as a template for messenger RNA

other strand of DNA

pore ← **3. messenger RNA passes out of the nucleus into the cytoplasm**

cytoplasm

ribosome

Ribosome

gly codon

ser codon

ser codon

anticodon

transfer RNA molecule which brings its amino acid to the messenger RNA molecule

4. at the ribsome, amino acids are assembled into proteins according to the code on the messenger RNA molecule

peptide link   peptide link

gly   ser   ser

protein being assembled

Figure 3.7 How the DNA molecule replicates – the two strands of the DNA molecule begin to separate and new nucleotides fall into place by base pairing

It is worth re-emphasising the point that DNA truly is the key to life because:

- ■ It controls the production of proteins.
- ■ All enzymes are proteins.
- ■ All living processes are controlled by enzymes.

Gene switches

The understanding of the genetic code by 1966 marked the end of an era in molecular biology and genetics. Perhaps one of the most important unsolved problems left was to find out what tells genes to work or not to work. It is illogical to think that genes are continually producing proteins whether they are needed or not.

Every cell in an animal or plant's body contains the same number of chromosomes according to the species. Each of these chromosomes carries a collection of genes in its DNA, and it is believed that the DNA is identical in each cell. Yet the cells themselves vary greatly in structure and above all in function.

After fertilisation, for the first few divisions of the new cell all the resulting cells are identical. Soon however something tells the cells to start to become the tissues and organs which they are destined to become. By the time a human embryo becomes recognisable as such, many different types of cells have formed. From an original blob of cytoplasm with a nucleus containing both parents' chromosomes arise nervous, muscular, digestive, excretory, circulatory, reproductive and respiratory organs. Furthermore, each type of cell within an organ uses its genes to a varying extent at different times, depending on its needs at that moment. For example, a liver cell may produce lots of one enzyme at one time and then switch to making another later on. The genes that are not being used are still there in the chromosomes but they are **repressed** or switched off until needed.

Something must determine whether a gene is repressed or not. **Feedback control** is a widespread phenomenon in biological systems and is responsible for the self-regulatory control mechanism which switches genes on or off. One familiar example of this type of control is a central heating thermostat. If we set the thermostat at a certain temperature, the heating system switches itself on as soon as the temperature drops below the predetermined value. In ourselves there are similar regulators. When we run, the heart speeds up because the muscles are using up more oxygen than usual. This is a signal that they need a better blood supply to bring them more oxygen and glucose. If we eat sweet or starchy food, this is digested to sugar molecules which are carried around the bloodstream. The rise in blood sugar concentration causes the pancreas to increase its production of insulin, a hormone needed to remove excess sugar from the blood.

These automatic feedback systems affect the body as a whole. Inside cells there are other feedback control systems which govern the functioning of the enzymes. Jacques Monod and Germaine Cohen-Bazire of the Pasteur Institute first discovered that a cell's synthesis of a particular enzyme is controlled by the concentration of the substance the enzyme produces. These French researchers were investigating the enzyme tryptophan synthetase, which catalyses the synthesis of the amino acid tryptophan. They found that when *Escherichia coli* bacteria were growing in a medium containing plenty of tryptophan, they stopped making tryptophan synthetase. This is an example of negative feedback, where the presence of a product helps stop the mechanism for making it.

A positive control is also needed in the cell so that it can start to make a substance in response to a new need. The first explanation of this mechanism began at the beginning of the twentieth century by Frederic Dienert of the Agronomical Institute in France. Dienert was investigating the way in which yeast cells ferment lactose, the sugar found in milk. He found that when strains of yeast had been grown in a medium containing lactose for several generations, their enzymes were able to deal with this sugar. Once given lactose they would start fermenting it within an hour. However, those yeast cells that had not been grown in lactose were not able to ferment it. Within fourteen hours though they would have adapted and would be making the necessary enzyme to ferment the sugar. It looked as though the presence of lactose had switched on the synthesis of the very enzyme that could destroy it by breaking it down.

Before an enzyme can be synthesised, the gene that controls its formation must go into action and send appropriate instructions via messenger RNA. Once this has happened the enzyme will be synthesised on the ribosomes automatically according to the specifications coded in the messenger RNA. The interesting point was how the action of the gene is switched on or off according to the cell's requirements. Some brilliant work on this problem was published in 1961 by François Jacob and Jacques Monod at the Pasteur Institute. They had been studying how gene mutations affected the ability of *Escherichia coli* to produce enzymes. When given a suitable sugar these bacteria, like yeast cells, make the necessary enzymes to break down the sugar. Jacob and Monod investigated the bacteria's production of beta-galactosidase, which is the enzyme that breaks down lactose sugar. They found that one kind of gene mutation stopped the bacteria from making the enzyme under any circumstances. They concluded that the affected gene was responsible for actually making the enzyme and called it a **structural gene**. Another type of gene mutation destroys the bacteria's usual ability to regulate the production of beta-galactosidase. Instead of making the enzyme to order only, the mutant bacteria pour it out all the time. Jacob and Monod concluded that the structural gene was normally under the control of a second **regulator gene**. The functioning of the regulator gene had been destroyed by this second mutation. Since the mutation of the regulator gene caused non-stop production of the enzyme, they deduced that the regulator would normally switch the structural gene off by producing some substance to inhibit the structural gene's activity.

In order to provide evidence for this prediction Jacob and Monod devised an ingenious experiment based on the knowledge previously gained by Jacob and another worker, Elie Wollman. They had found that some strains of *E. coli* can 'mate' by passing chromosomal material from the male cell to the female cell. The material is passed between the cells like a rope. In a crowded attic laboratory, Jacob used his wife's kitchen blender to whiz up the bacteria as they mated. Jacob and Wollman found that after breaking off the bacteria's sexual relationship in this way, only part of the rope had been inserted. Interrupting mating at different intervals gave the times for different lengths of chromosomal material to be passed across. It took eighteen minutes for four genes to pass from the male to the female, twenty-five minutes for five genes, thirty minutes for six genes, and so on.

Using the same method, Jacob and Monod now investigated how the regulator gene relates to the structural gene. They used a female strain of *E. coli* in which both types of gene concerned with the production of beta-galactosidase were defective, and a male strain in which both genes were working normally. To begin with both male and female strains were grown in the absence of lactose. Then the two strains were mixed together and mating began. From previous work it was known that both the regulator and structural genes passed into the female about eighteen minutes after mating had begun. Soon after the genes had passed into the female cells, they began making beta-galactosidase, even though there was no lactose present. They were acting like mutant cells in which the regulator gene was destroyed. About an hour later, however, the production of the enzyme stopped – the gene was switched off. Jacob and Monod reasoned that at this time the regulator gene had gone into action and made some chemical which stopped the structural gene from making any more beta-galactosidase. For this research they were awarded a share of the Nobel prize for physiology and medicine in 1965.

Yet another type of mutation affects the production of beta-galactosidase. This third type prevents any switching off of the structural gene even though the regulator gene is intact. Jacob and Monod concluded that there must be a third type of gene, which they called the **operator**, that controls the functioning of the structural gene or of a group of structural genes. Fine genetic analysis has shown that the operator gene is next to the structural gene that it controls. It appears that the repressor substance made by the regulator gene does not act directly on the structural gene but it binds itself to the operator gene.

An operator gene together with its structural genes forms a functioning unit which Jacob and Monod called an **operon**. Regulator genes can control one or more of these operons. The regulation of an operon depends on the presence of some outside chemical in two ways, the inducible and the repressible. The inducible system is an example of positive feedback. The represser substance made by the regulator gene interacts with the outside substance such as lactose. This prevents the represser substance from reaching the operon and repressing it. As a consequence, the operon sends out instructions to make the enzyme needed to break down lactose. The repressible system is an example of negative feedback. Here the represser is not effective until it has interacted with the product of the system which the operon controls. For instance, an essential component in the manufacture of the amino acid tryptophan is the enzyme tryptophan synthetase. This enzyme is produced automatically according to instructions from the operon until the represser interacts with some of the tryptophan produced. Then the represser becomes effective and inhibits the functioning of the operon, stopping the enzyme production. Thus the amount of tryptophan produced is regulated by the amount already present.

4 | GENETICS AND SOCIETY

Mutations and history

Haemophilia – a mutation in Queen Victoria

A mutation in a sex chromosome of Queen Victoria influenced the fates of both the Russian and Spanish royal families to such an extent that the repercussions echoed around the whole of Europe. In Chapter 2 we learned about sex linkage and certain mutant alleles which occur only on the X chromosome. One such allele leads to a condition called **haemophilia**, the inability of the blood to clot properly.

The blood-clotting mechanism is a complex series of enzyme-controlled reactions involving a sequence of intermediate chemicals leading to the formation of a clot to seal over a cut. Any defect in this chain of events leading to the formation of the blood clot can result in haemophilia, a potentially life-threatening condition in which a person may bleed to death following even slight damage to the blood vessels. There are several varieties of haemophilia caused by abnormalities in one or more links in the chain. The most common is a defect in the blood protein Factor VIII, which results in classic haemophilia or haemophilia A. The other main form of the disease, Christmas disease or haemophilia B, is due to a defect in Factor IX. Both forms are inherited in the same way, as a recessive sex-linked allele on the X chromosome.

Women who are carriers of haemophilia A have one normal dominant allele for the production of Factor VIII and one defective allele. They produce less than the normal amount of Factor VIII but this does not seem to cause them problems. Because a male has only one X chromosome, which originates from the mother, the presence of the mutant allele means he will suffer from haemophilia. A female will only suffer from the disease if she has the mutant allele on both her X chromosomes, which is very unlikely. Statistically then, half the sons of a carrier will suffer from haemophilia and half the daughters will be carriers – they will be heterozygous for the condition.

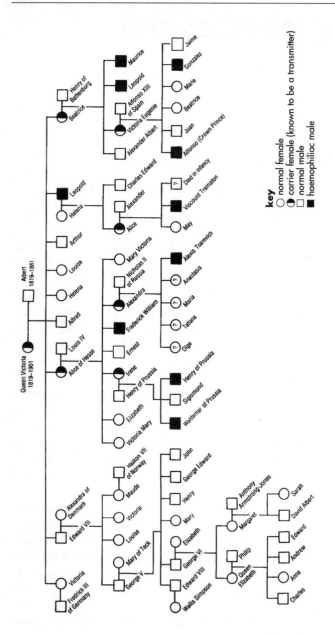

Figure 4.1 The inheritance of haemophilia in the descendants of Queen Victoria

You can see from Figure 4.1 that three out of four of Queen Victoria's sons were very fortunate not to have inherited the disease. Even among her five daughters only two were carriers. This illustrates the limited value of statistical probabilities in small samples. The reasons for using prolific organisms like bacteria, peas and fruit flies in classic genetical experiments may now be clearer!

All the daughters of haemophiliac men must be carriers, because they inherit the allele on the X chromosome from their fathers, but all the sons of haemophiliac men and their subsequent descendants will be normal. Figure 4.1 shows how the mutant allele in Queen Victoria spread throughout the pedigree.

The Russian royal family

The significance of this pedigree to the Russian royals began with Alexandra's marriage to Tsar Nicholas II. Alexandra was a carrier of the mutant allele, having obtained it via Princess Alice, one of Queen Victoria's daughters. When Alexandra became the tsarina she was to provide the Russian hereditary rulers with a male heir to the throne. As you can see from the pedigree her first four children were daughters. However, child number five was born on 12 August 1904, and the guns of St Petersburg boomed out a 300-gun salute – it was a boy! Tzar Nicholas was so delighted that he chose that day to pronounce the immediate abolishment of the barbaric corporal punishment in the Russian armed forces which was common practice at the time.

For a few weeks all seemed well with little Alexis but by September they noticed persistent bleeding from the child's navel. At first they were reluctant to admit that the child might have inherited the 'bleeding' condition, but eventually they had to face facts. His mother realised that the affliction was transmitted by the female line and that it was common in her family. However, the Russian people considered any defect a divine judgement for some sin. The tsar was head of the Church and as such was expected to be perfect. The defect in the future tsar had to be concealed at all costs from all except close family members and some courtiers who were sworn to secrecy. The depressed Alexandra turned all her thoughts to religion and the secret forced the family even closer together in unhealthy isolation.

Then in 1905 the now infamous Grigori Rasputin came on the scene. It would be difficult to imagine a more unlikely character mixing with royalty. Rasputin was an alleged Siberian horse thief of peasant stock, with an ugly scarred and gnarled face and a filthy beard. This man came to have a major influence in the fate of the royal family of one of the world's greatest nations. Rasputin believed in a religious cult which preached the bizarre dogma of salvation through sin, especially to female followers. His teachings stated: 'Man must sin in order to have something to repent. If God sends temptation, we must yield to it voluntarily and without resistance so we may afterwards do penance in our contrition.' He provided the temptation, followed by many opportunities to sin, and then gave his own personal absolution!

There is much recorded evidence of his hypnotic powers. He used to pacify Alexis when he became hysterical during bouts of bleeding, sending him to sleep with consequent lowering of his blood pressure. This was probably the only therapy that could stabilise the boy's condition, and it obviously impressed the tsarina. It is possible that Rasputin's hypnotic abilities could have constricted the blood vessels, thereby reducing bleeding. There were several times when the tsarina commanded the presence of Rasputin when Alexis appeared to be near death. Accounts by members of the family describe almost miraculous recoveries as a result of Rasputin's prayers.

The tsar and tsarina became so obsessed by Rasputin's powers to provide comfort for their son that they granted him a degree of protection normally more suited to royalty. He was given lavish apartments which became dens of iniquity for rituals connected to his unorthodox cult, and his debauchery and decadence were tolerated by the tsar. Many state decisions made by the tsar were influenced by Rasputin, and the public lost all respect for the tsar.

By the end of the First World War many in the Russian government saw that their country would be doomed if Rasputin was left to influence future developments. They plotted to murder him and fed him with cakes laced in potassium cyanide, shot him four times and pushed him under the ice in a frozen canal. An autopsy showed he was still alive when put into the frozen canal!

The failure of the cyanide to kill him quickly suggested that Rasputin himself was a mutant and did not secrete hydrochloric acid from his stomach glands. In most of us the hydrochloric acid would react with the cyanide to produce deadly hydrocyanic acid.

Rasputin had contributed to the destruction of the credibility of the Russian monarchy, and the scene was now set for the communist revolution which began in 1917, shaping the destiny of millions. Rasputin's influence had all come about because of a mutant allele in Queen Victoria's X chromosome!

The Spanish royal family

The very same chromosome also took part in shaping the history of Spain. Here a succession of physically defective heirs helped to discredit the liberal forces who had encouraged the King to take a British bride. This was a significant factor in the events leading to the Spanish Civil War. Queen Victoria's youngest daughter Beatrice was a carrier of haemophilia and transmitted the gene to three of her four children, one of whom, Eugenie, became Queen of Spain.

The King's choice of bride was of considerable political importance. Alfonso favoured a British bride although he knew that a Roman Catholic convert would be an unpopular choice for queen in some Spanish circles. The fact that her brother had haemophilia made it likely that Eugenie was a carrier. Alfonso was warned of the risk but ignored the potential problem. The marriage drew Spain towards England and the new queen became the social focus of Liberal society.

As can be seen from Figure 4.1 just two of the male heirs were not afflicted by the condition. One of these, Jaime, was born a deaf mute. It is difficult to measure the effect of these tragedies on Spanish royalty. At the beginning of the twentieth century Spain was so torn apart by fanaticism and extremism on both left and right that civil war may have been inevitable. However, these genetic misfortunes undoubtedly contributed to the weakening of the position of the throne and led to the mistrust of Eugenie's English background by true blue-blooded Spaniards.

The source of the problem

What was the origin of Queen Victoria's haemophiliac allele that had such devastating and far-reaching effects? There are two possibilities – either the mutation originated in Queen Victoria, or Victoria's father was not the Duke of Kent. None of Victoria's ancestors for many generations show any evidence of having had haemophilia. Most haemophiliacs inherit the condition from one of their parents but a proportion result from a new mutation. The mutation is unlikely to have been present in Victoria's mother, who had a son and daughter by her first marriage whose large number of descendants show no sign of haemophilia. It is possible that the mutation occurred in the Duke of Kent, but it is astonishing that another genetic defect possessed by the Duke was not passed on to Victoria or her family. This genetic defect was porphyria, the calamitous scourge of George III, father of the Duke of Kent.

The madness of George III

Another mutation that affected the course of history was the debilitating condition of porphyria suffered by George III (1738–1820). This caused the 'madness of George III' and among other things led to the establishment of the science of psychiatry, then called 'the mad business'. Ironically it has now become clear nearly 200 years after his death that George III was by no means psychotic. The much discredited and maligned king suffered spells of pain and delirium from a metabolic disorder that was not recognised until the twentieth century as being porphyria. His illness precipitated the notorious Regency Crisis in which the Prime Minister of the day William Pitt was in danger of being expelled. It was also popularly believed that during one of his bouts of illness the king made decisions which sowed the seeds of the American War of Independence and thus contributed to the loss of the American colonies.

One of the symptoms of porphyria was observed by George III's physicians, although its significance was not realised. This was the king's dark red urine. The coloration was caused by a red chemical called porphyrin, a pigment present in the haemoglobin of the blood. Normally porphyrin is broken down in the body cells by an enzyme, so its presence in the urine is a sign that this critical enzyme is missing, the result of a mutant gene.

The clinical seriousness of a very similar defect was first brought to the attention of the medical world in 1908 by the London physician Sir Archibald Garrod (see page 96). He was studying a rare inherited disorder called alcaptonuria, in which the urine turns black and smelly after eating certain foods. He found that the smelly substance was a chemical resulting from a failure to complete the breakdown of a certain substance in the food. Garrod predicted that the symptoms of alcaptonuria arose because an enzyme was not working properly.

The more serious problem of porphyria affects the nervous system. Attacks begin in the nerves and eventually reach the brain, resulting in paralysis, delirium and agonising pain – not a desirable state in which to make world-shaping decisions!

The king's inheritance of porphyria can be traced as far back as his sixteenth-century ancestor Mary Queen of Scots. In 1786 the disease claimed Frederick the Great of Prussia as a victim. Edward Duke of Kent,

Queen Victoria's father, suffered severely from the condition and died of an attack a week before the death of George III. The condition persisted in all the descendants of George III but curiously not in Queen Victoria.

The human genome project

Since the 1980s advances in the fields of both genetics and medicine have led to developments in medical genetics, evolving medical practice at a rapid pace. The **human genome project** is a global research effort that aims to analyse the structure of human DNA and map the location of an estimated 100 000 genes. The maps of newly discovered countries made by the first explorers were arguably the first steps to exploiting the territories they revealed. In the same way the human race must protect the definitive map of human genes from exploitation by the unethical. It is anticipated that the end product of this research will be the standard reference for biomedical science in the twenty-first century and will help us understand and eventually treat many of the four thousand plus recognised human genetic afflictions. The goals to be achieved are:

- mapping and sequencing the human **genome** (all the human genes)
- mapping and sequencing the genomes of certain other organisms
- data collection and distribution
- consideration of the ethical and legal implications
- research training
- international sharing of ideas in gene technology.

Initial estimates suggest all this will take up to fifteen years. The aim of the enormous research programme is to sequence the quartet of chemicals adenine, guanine, cytosine and thymine in human DNA. The four bases making up DNA (see page 107) are represented no less than three thousand million times in humans. If typed in order the bases of the human genome would fill the equivalent of 134 complete sets of the *Encyclopaedia Britannica*. The size of an individual gene within the whole length of human DNA is similar in comparison to the size of an ant on Mount Everest! This mind-boggling project will provide a reference for medical doctors and scientists who wish to study the genetical basis of inherited disorders, human development and evolution.

The implications for society

The public's general interest in the human genome is primarily focused on serious genetic disorders within families or among friends. However, there are many other advantages from this research. Medical science has made great progress in understanding the fundamental basis of cancer over the past fifteen years or so, most of which originated from the discovery of cancer-causing genes (oncogenes) and now also anti-oncogenes. Progress on this scale depends on the study of cancer at DNA level. Gene research can have an equally important impact on a number of other disorders. The actual cause of Alzheimer's disease is still a mystery, although it is known that some families seem to have a genetic predisposition to it. Recent studies confirm that a gene on chromosome 21 is responsible for a minority of Alzheimer's cases.

Until the late 1970s clear human genetic maps were impossible even to contemplate. By 1985 extensive genetic maps of most human chromosomes became available because of improved techniques of DNA analysis. However, there are many moral and social questions associated with this knowledge. The examination of people's individual genes may show some that are imperfect. Once this has been discovered, will the lives of those with mutated genes be made worse? Should other people have access to information about our genes? Insurance companies are certainly interested in your genes when considering life assurance policies. If it was known, for example, that a person had the gene for Alzheimer's disease, the attitude might be: 'You are probably going to be senile at the age of fifty-five and so we are not going to employ you or let you have insurance.' Some people might say that there should be legislation preventing the screening of a person's genome for harmful genes. However, to block the whole genome programme because of fears that it might be misused would prevent the treatment of a wide range of diseases and disorders. Perhaps we should go ahead and find the genes, but ensure that the information is not divulged to potential misusers. These sort of ethical dilemmas will need to be resolved in parallel with the ongoing progress in medical genetics.

Genetic screening

In 1997 the first recommendation for any nationwide genetic test was made in the United States by the National Institute of Health. Recommendations of NIH panels are not mandatory but they are generally adopted by doctors. This recommendation said that all couples planning to have children should be offered a genetic test for cystic fibrosis (page 34). About 800 children are born with cystic fibrosis in the USA each year. With an incidence of one in 30 people carrying the recessive gene, it is the most common inherited disease in the USA. There is a one in 900 chance that an average couple will both carry the mutant gene and if they do there is a one in four chance that their child will have the disease.

The information from screening could be used by couples to decide whether to risk having a child in the first place, whether to go ahead but test the fetus with a view to possible termination of the pregnancy if it is affected, or whether to complete the pregnancy nevertheless but prepare for the possibility of the child suffering from cystic fibrosis. However, the NIH panels do not suggest routine screening of newborn babies because there is no point in starting treatment before symptoms show, usually at about six months.

Eugenics – the ethics of interference

Eugenics is the theory that the human race could be improved by controlled selective breeding using individuals with desirable characteristics. Apart from the ethical considerations of whether this is desirable in the first place, there are practical problems with the judgement of what are desirable characteristics and also with the problem of assessing the role of genetic and environmental influences on development.

The road towards our present knowledge of human genetics is littered with debris from perverse and biased minds intent on improving the human race. There have been many people certain that they knew what was 'best' with regard to racial qualities and who firmly believed that they should decide which genes should survive in the human race. The earliest ideas of improving society by selective breeding began with Plato in 427 BC. He suggested a system of breeding festivals for gifted men and women with state-controlled nurseries for the offspring. In Aristotle's *Politics* (384 BC), abortion by choice is advocated along with the humane killing of the mentally and physically handicapped. It took another two thousand years before the mechanism of genetics became regarded as a science but even in historically recent times, the thoughts of a significant number of influential people have been directed along very similar pathways.

One of the first of these influential dogmatic thinkers was Francis Galton (1822–1911), a first cousin of Charles Darwin. It was he who invented the term eugenics in 1883. In his book *Hereditary Genius* (1869), he advocated judicious marriages between men of genius and women of wealth. Galton was a wealthy Victorian who had qualified as a medical doctor but never practised medicine. His views were blatantly racist, sexist and elitist and he imposed his imagined superiority on all who had the misfortune to meet him, particularly during his travels in Africa. However, he was considered to be a genius ahead of his time, and his positive contribution to science was his work in biometrics, the application of statistical analysis to biological data. He was one of the first to recognise that identical twins were genetically the same, but he underestimated the effects of the environment on development. Before his death he endowed a research post in eugenics at University College, London, and founded the Eugenics Society.

Another of the same era was George Bernard Shaw (1856–1950). The following quotation shows that 2000 years of civilisation had not touched the prejudices of a mind considered to be one of the finest of its day: 'If we desire a certain type of civilisation, we must exterminate the sort of people who do not fit into it.'

Probably the best known scientist to take the Galton research post in eugenics at University College, London was professor Karl Pearson, a brilliant mathematical geneticist. The extreme racial views of inheritance which spread throughout Europe and the USA in the first half of the twentieth century probably stemmed from his work on eugenics. Scientists with preconceived ideas began falsifying data and many insinuations regarding the inheritance of intelligence and criminality were erroneously made. Eugenics was overtly practised in the USA in the 1920s and 1930s – by 1931 most states had sexual sterilisation laws aimed at eliminating the mentally retarded, epileptics and sexual deviants. Over 1000 people were sterilised under these laws in California. At the same time the belief in the superiority of the Aryan race was growing among certain fanatics in Nazi Germany leading to the most horrific holocaust that the world has ever known. Two quotations from totally opposed leaders of that time are interesting:

'It is outstandingly evident from history that when the Aryan has mixed his blood with that of inferior peoples the result of mixed breeding has invariably been the ruin of the civilising races.' (Adolf Hitler)

'The unnatural and increasingly rapid growth of the feeble-minded and insane classes constitutes a national and race danger which is impossible to exaggerate. I feel that the source from which the stream of madness is fed should be cut off and sealed off before another year is past.' (Winston Spencer Churchill)

Hitler believed that 'best' meant Aryan – tall, slim, blond, athletic, blue-eyed and intelligent with leadership qualities. He thought that only the people who had these qualities deserved to rule the world. He also believed that the master race should be pure bred, the genetic equivalent of Mendel's green and wrinkled peas!

In 1936 the Olympic Games were held in Berlin and the world famous black American Jesse Owens won four gold medals. Hitler was furious because this proved that his ideas about Aryan supremacy were total nonsense. Hitler refused to present Owens with his medals.

The fact that the leaders of the Nazi party showed no pure Aryan qualities themselves did not seem to concern Hitler. He and his henchmen Goebbels, Goering and Mussolini could hardly be described as Aryans. When the Japanese joined forces with the Nazis, some serious thinking had to be done. Logically the superior Aryans should have considered the Orientals as inferior. It was decided that the Japanese had a sufficient historical contact with Aryan people to have interbred significantly. Although the Japanese looked oriental, it was said that nevertheless they possessed all the moral and intellectual qualities of an Aryan. It was then officially announced that the Japanese leaders were as biologically reliable as the Germans. The nonsense of such thinking would be comical if it were not for the consequences of the Second World War. As for the quotation from Churchill, perhaps we can be thankful for democracy terminating such ideas while they are still embryonic.

Has the human race learned any lessons from these eugenic disasters? In 1995 the Chinese government passed a controversial eugenics law which requires couples who want to marry to undergo screening for inappropriate genes. Marriages are only allowed if couples with such genes agree to sterilisation or long-term contraception. The loose wording of the law gives the government a free hand to decide which genes to test for. China's population of 1.2 billion people offers a wealth of genetic information. The country's rural millions have remained relatively static in the twentieth century, so each region has a unique blend of genes and diseases. This makes it easier to trace pedigrees of inherited disorders back to defective genes. Can we be sure that information gained from this massive screening will not be used against the people involved in it?

Hello Dolly!

The sort of eugenics described above has at least involved the normal reproductive process of fertilisation. Early in 1997 genetics entered a new era with the arrival of Dolly the sheep. As a product of cloning rather than of fertilisation she was the most remarkable mammal ever to be born. A **clone** is a group of identical organisms descended from a single ancestor, so having the same DNA.

Dolly was created from a cell taken from the udder of a six-year-old ewe. Proving that scientists do have a sense of humour after all, albeit one with a sexist bent, she was called Dolly after a world famous country and western singer who happened to be well blessed with mammary gland tissue.

The cloning of sheep from different cells of the same embryo had been done previously in 1996. The difference with Dolly was that all her DNA came from one cell of an adult sheep. The work was done at the Roslin Institute in Edinburgh, and if it were applied to humans it could mean that each of us could have clones of ourselves made from our own tissue, containing the same DNA as us.

To produce Dolly the Roslin team of researchers combined material from two sources. First they took immature unfertilised egg cells called oocytes from the ovaries of ewes. They removed the chromosomes from these, leaving DNA-free cells. Next they took cells from the udder of a mature ewe and fused the empty oocytes with the udder cells by subjecting them to an electric current. After culturing these dividing cells for a week, one of the fused cells formed an early embryo called a blastocyst. This was implanted in the womb of a surrogate mother. Five months later Dolly was born. Ian Wilmut of the Roslin team recorded that it was a very hit or miss process, with a success rate of one in 277 attempts at fusion. He suggested that the failures were due to the difficulty in ensuring that the empty oocyte and the donor cell from the udder were at the same stage of the cell division cycle. Figure 4.2 shows the cloning process.

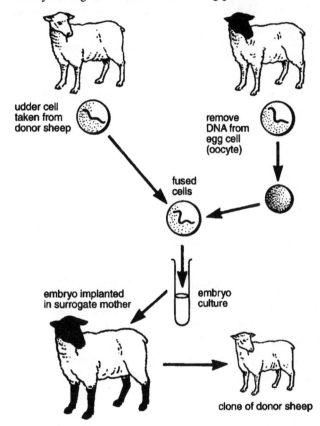

udder cell
taken from
donor sheep

remove
DNA from
egg cell
(oocyte)

fused
cells

embryo implanted
in surrogate mother

embryo
culture

clone of donor sheep

Figure 4.2 The cloning of a sheep to produce Dolly

The first pioneers in this field of research worked with frogs in the 1970s. A team led by John Gurdon at the University of Cambridge transplanted nuclei from the skin cells of adult frogs into frog eggs which lacked nuclei. Some embryos grew into tadpoles, but none metamorphosed into adults. Wilmut attributed his success with Dolly to the unique way his team treated the donor cells. They placed the cells in a salt solution and induced a state of hibernation by starving the cells, providing just enough growth factor to keep the cells alive. The cells stopped dividing and copying DNA, and seemed to switch off all but the most important genes. It was this treatment that enabled the nucleus-free egg to reprogram the donor nucleus.

The Roslin team's work is directed towards agricultural applications. Although there are some economic advantages of having stocks of genetically identical animals, traditional animal breeding makes progress because of genetic variation, whereas cloning fixes the genome – this eliminates new problems but also prevents new advantageous features emerging.

Of course it may be theoretically possible to carry out the same process with any mammal including humans, but most would consider this unethical and it is certainly illegal in Britain. However, in 1997 human cloning was still legal elsewhere including the USA. Today the implications of human cloning seem like material for a science fiction horror story. It is said that the leader of the Roslin research team received requests to resurrect dead relatives and a beloved pet. Many theologians are totally against the idea of humans interfering with nature and producing a person from a single cell, in spite of the idea of Eve being made from a single rib of Adam!

Since 1984 the prospect of cloning humans has been a major concern in submissions to Britain's Warnock Committee, which is concerned with the implications of genetic research. The possibility of creating, for example, brain-dead copies of humans as sources of perfectly matched organ transplants may be shocking at present, but if anything is theoretically possible, given sufficient time it could happen somewhere.

5 | GENE TECHNOLOGY

Genetic engineering

The science of genetics is undergoing a revolution, largely because of a marriage between genetic engineers and multinational companies with profit in mind. The result is a battery of living factories which can produce proteins on demand. The living factories are yeast or bacteria cells and other microbes modified and harnessed to work for us.

An outline of genetic engineering

The term **genetic engineering** is familiar to most people today, having been used in both emotive and trivial contexts by the media and also by science fiction writers. Genetic engineering embraces many concepts including gene manipulation, gene cloning, recombinant DNA technology, gene therapy and genetic modification. In a nutshell, genetic engineering means finding specific genes, cutting them out of chromosomes and splicing them into chromosomes of other species. After the genes have replicated many times, the proteins made by them are harvested.

The main advantage of this technique is the mass-production of many otherwise scarce and expensive protein-based chemicals which have a direct role in saving lives. For example, the extraction of just 5 mg of the protein somatotrophin (a growth-regulating hormone) would require a half a million sheep's brains! The same mass of this hormone can be made in one week by nine litres of genetically engineered bacteria.

The genetic engineer uses some highly specialised equipment and terminology:

■ The **source DNA** contains the required gene, which is cut out and added to the host DNA.

■ The **host DNA** is cut to allow the insertion of the source DNA.

■ The **recombinant DNA** is hybrid DNA resulting from the fusion of the source DNA and the host DNA.

■ **Restriction enzyme** is an enzyme which cuts DNA at specific points, allowing the source DNA to be inserted into the host DNA.

Isolating a gene

If an organism produces a particular protein, then you know that the organism must have the gene which codes for the protein's production. The skill comes in searching through its chromosomes and finding exactly where the gene is positioned so that it can be cut out and replicated.

One approach is to treat the cells containing the gene with restriction enzymes. These will chop up the chromosomes, cutting them in specific places and leaving bits that can be inserted into the DNA of host chromosomes. In this way a collection of cells is produced, some of which will carry the desired gene in their chromosomes. The geneticist must isolate those cells that contain the required gene. This can be done by identifying the cells that make the protein – some proteins interact with certain chemicals called antibodies, and if such a reaction is detected the gene must be present. Another way of finding a particular gene is to expose its complementary RNA on a column and flush the gene fragments down the column.

Cutting and splicing the DNA

Bacteria normally make one or more restriction enzymes to defend themselves against attack by viruses. When a virus invades a living cell, it takes over the metabolic activities of the cell and begins reproducing itself (see page 87). The newly formed viruses burst out of the cell when it ruptures and rapidly attack other healthy cells.

There are strands of DNA or RNA inside viruses, coated with a protein layer that protects the nucleic acid from most of the cell's enzymes. However, restriction enzymes produced by the bacteria can penetrate this line of defence and will chop up the viral DNA. This is the origin of the name restriction enzyme, because it restricts the attack of bacteria by viruses.

Different restriction enzymes work at very specific places along the DNA strand. The type made by *Eschericia coli*, bacteria which naturally live in our intestine, recognises the following sequence of DNA bases:

```
GAATTC
```

and its complementary sequence:

```
CTTAAG
```

The enzyme breaks the DNA wherever these sequences appear, in the following position:

```
G        AATTC
CTTAA        G
```

The result is two DNA strands with uneven 'sticky ends', that is, they tend to stick to other molecules with a complementary sequence of bases. Restriction enzymes do not distinguish the source of DNA – they will cut the specific sequence of bases wherever it exists, and whenever complementary 'sticky ends' meet, they join.

After disrupting the source cell and subjecting the chromosomes to a specific restriction enzyme, the genetic engineer is left with pieces of source DNA with specific 'sticky ends' containing a particular sequence of nucleotides. The next step is to insert these fragments into bacteria, where they can be cloned. *E. coli* is one of the most common bacteria used for cloning, and like many other bacteria it has a large circular chromosome and also a small circle of DNA called a plasmid, as shown in Figure 5.1. Plasmid rings can be cut open with the same restriction enzymes used to isolate the fragment of source DNA. When the source fragments and the cut plasmids are mixed together, the appropriate complementary 'sticky ends' join up in the presence of another enzyme called a **ligase**. The result of this splicing is recombinant DNA, a plasmid containing the source fragment, which must be inserted into the host bacterium. The bacteria are mixed with calcium salts which make their cell walls permeable to the altered plasmids. Once inside the host cell, the new plasmids replicate. Their new DNA sequence is also reproduced each time the host cell reproduces. In this way the recombinant DNA is cloned. The host bacteria are allowed to reproduce in ideal conditions and when there are sufficient numbers of bacteria, they are broken open and the protein can be harvested.

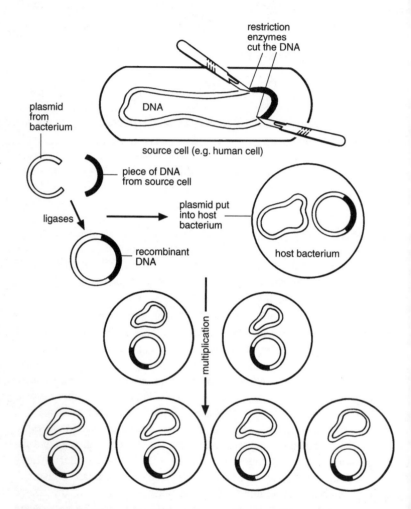

Figure 5.1 The principle of genetic engineering

Separating the useful from the useless

After splicing and cloning genes, the genetic engineer has a population of bacteria, some of which contain plasmids and some which do not. The problem is to find the cells that contain plasmids and separate them from the rest. This is not as difficult as it might sound because most plasmids used in cloning give antibiotic resistance to their hosts. The host population is subjected to antibiotics and those that die are the ones without the plasmids. The rest are cloned.

However, the difficult part is yet to come. Remember that the restriction enzymes cut the source DNA into fragments, chopping it wherever a particular nucleotide sequence occurred. Various different fragments could have been incorporated into a plasmid because they all had the appropriate 'sticky ends'. The result is that some plasmids have the desired gene and some do not. How do you separate those cells carrying plasmids with the gene that you want from the rest? The process is complex, involving screening for the protein produced by the gene of interest. The nucleotide sequence of the gene can sometimes be deduced by analysing the amino acid sequence of the protein for which it is responsible. If the nucleotide sequence is known, nucleic acid segments can be chemically synthesised and used to identify the gene by bonding to its complementary sequences. Today most laboratories specialising in this type of research carry such nucleic acid sequences in stock and the molecules can simply be made to order. This is called cloning by 'phone.

The potential for mis-use

The mass-production of life-saving proteins through genetic engineering is already with us, but what about the potential life-threatening products of gene manipulation? For example, what would happen if someone inserted a cancer-causing gene into the familiar *E. coli* that lives in our intestines, or a gene that makes a deadly toxin? What if these bacteria were then released into a human population? This may seem unlikely, but history books now tell us about mustard gas of the First World War, the atom bombs of the Second World War, napalm in the Vietnam war and nerve gases of other recent wars – someone invented them and someone used them! Biological warfare is not well documented in popular science publications because most developments are classified as top secret, but behind the closed doors of secrecy all is possible.

Another less cynical concern about genetic engineering is that well-meaning scientists could accidentally allow a potentially dangerous genetically engineered organism to escape into the environment. Even after smallpox was eradicated from Earth, there were two minor epidemics in Europe caused by cultured experimental viruses that had escaped from a laboratory. One person died of a disease that technically did not exist!

Genetic engineers have always been aware of this accident waiting to happen and have safeguarded against it. In the same way that they can produce recombinant DNA which carries genes for manufacturing useful products, they can introduce genes into that same DNA which make it possible for the microbes to survive only in highly specific conditions. If these altered microbes were to escape they would die. In the early days of gene manipulation using *E. coli*, a mutant was produced that had a defective gene preventing it from making its usual protective cell coating. The material that the gene normally made had to be provided artificially in the culture medium. However, you will be aware by now that microbes mutate naturally, on their own. There is a biological rule which says that, given a sufficiently large population and sufficient time, organisms will adapt to pressures of environmental changes in such a way that it is to their advantage. In this instance, the microbes soon mutated back to the form that could produce the coat. Another gene responsible for the production of the coat was deleted. Even this was not enough; they reproduced anyway.

The pioneering scientists investigating the problem were two Americans, Roy Curtiss III and Dennis Pereira. The latter found that the microbes were manufacturing a sticky substance called colanic acid which acted as a protective coat. When they destroyed their ability to produce this chemical by altering the offending gene, they finally had a tailor-made bacterium which could not exist outside a highly specialised human-made environment. Unexpected bonuses are often a product of research and this was no exception – the new microbe was also sensitive to ultraviolet light, so would be killed in sunlight.

One problem remained. About one in a million bacteria is sexually active and reproduces by mating (conjugating) with another. So, theoretically, if an escapee mutant *E. coli* mated with a normal bacterium any dangerous gene could be passed on to found a colony of unwanted mutants. Curtiss solved the problem by altering yet another gene in the mutant, one which normally controls the manufacture of thymine, an essential constituent of DNA. Thymine therefore must be supplied in the culture medium. The microbe was finally crippled and rendered totally dependent on laboratory conditions.

The implications of the new technology – threats or promises?

It is interesting to look back at the attitudes of the 1970s concerning the then new science of genetic engineering. In February 1975 scientists met in Asilomar, California, and held a heated discussion to draw up guidelines which would prevent the nightmare situation of genetically engineered mutants escaping. Rules were formulated and some communities banned research on recombinant DNA altogether. Later, in the 1980s, attitudes changed and much of the opposition to research and development in this area dissolved. Was it a coincidence that this acceptance occurred in parallel with the realisation of massive commercial feasibility in this field?

There are people who still fear the results of gene splicing and gene manipulation but no one can deny the promise of the technique to mass-produce life-saving chemicals in the next millennium. This line of research started in small laboratories often staffed by graduate students focused on a brave new future. Multinational drug companies then reacted with great enthusiasm, making stock brokers instant millionaires and tempting many a dedicated scientist to leave the ivory towers of academia.

Many possibilities emerged for recombinant DNA techniques, not the least of which was the potential for mass-production of hormones. Insulin, for example, could be obtained only from animal tissues and in relatively small amounts in the past. Today almost unlimited quantities of insulin can be harvested from genetically engineered microbes, changing the lives of many diabetics throughout the world. Other therapeutic drugs manufactured in the same way are listed in Table 5.1.

Genetic profiling

The uniqueness of your DNA

The fact that the sequence of nucleotides in our DNA is as individual as our fingerprints is the basis of a new method of identification which can be used in criminal cases. This so-called **genetic profiling** was first used in evidence in the late 1980s to establish the guilt of a murderer. Since then the technology has been the subject of intense scrutiny in courts of law and in the media. The term genetic fingerprinting was soon coined to

Table 5.1. Some products of genetic engineering technology

Product	Function
Human growth hormone	Promotes growth in children who are unable to produce enough of the hormone themselves
Interferon	Helps cells resist viruses
Interleukin	Stimulates the production of white blood cells that take part in the immune response
Insulin	Treats diabetics by enabling the body to regulate blood sugar
Renin inhibitor	Decreases blood pressure

describe its use in criminology, but patterns on the skin of the fingers have nothing to do with it! The analogy is the uniqueness of both a set of fingerprints and a DNA profile. The technique is particularly good for identifying culprits of violent crimes where body fluids such as blood or any other tissue are left at the scene of the crime. There is now a computerised database holding DNA profiles of all convicted criminals and this promises to be the greatest breakthrough in criminal justice since the discovery that fingerprints are unique to individuals. Forensic science laboratories have allocated significant resources to its further development.

The origins of genetic fingerprinting

It has been claimed that the ancient Chinese were aware of fingerprints and used a kind of fingerprint system thousands of years ago. However, it was an English assistant commissioner of the Metropolitan Police force, Sir Edward Henry, who invented a standardised means of classifying human fingerprints which has been used throughout most of the world since 1901. The use of genetic markers in the form of blood groups by forensic scientists began around the same time, after the discovery of blood grouping techniques by Karl Landsteiner (see page 63). However, to successfully identify the blood group there had to be a sufficient quantity of blood that had not degraded. Crimes were often left unsolved because of the uncertainty of the analysis of a poor sample of material.

A breakthrough came in the mid-1980s when Professor Alec Jeffreys at Leicester University showed the presence of many variable regions of DNA which did not code for amino acids. These regions were called minisatellites and there are thousands of them scattered throughout the chromosomes, probably having evolved as mistakes during the replication of DNA. The number of times that these regions are repeated gives individuality to the profile. If enough regions of DNA are examined, it is possible to obtain a genetic profile which is almost unique to an individual.

The technique is shown in Figure 5.2 and can be summarised as follows:

1 Cells with nuclei are obtained from blood or other tissue.

2 DNA is extracted from the nuclei of the cells.

3 The DNA is cut into fragments using restriction enzymes obtained from bacteria.

4 The fragments are separated according to size using the technique of electrophoresis (see page 100).

5 The fragments are transferred to a nylon membrane in a process called Southern blotting. The nylon membrane is sandwiched between the electrophoresis gel and sheets of blotting paper and the DNA fragments move across into the membrane.

6 Sodium hydroxide is added to the membrane which splits the DNA into single strands, leaving the base pattern intact.

7 The DNA fragments are identified with a **DNA probe**. This is a portion of DNA with a nucleotide sequence that is complementary to that of the minisatellite repeat unit. The probe is labelled with either a radioactive or chemiluminescent tracer which affects X-ray film.

8 X-ray film is placed over the membrane and the film is developed. The result is a pattern of bands something like a shop bar code showing where the probe has bound to the DNA.

This basic technique has been refined and since 1988 it has been reliable in crime detection.

In the 1990s a more sophisticated technique for DNA profiling was developed. The new technology required only very small traces of DNA and the reactions in the process happened quickly, reducing the time taken

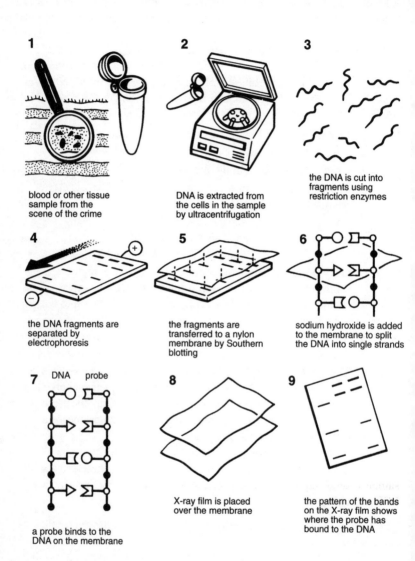

1

blood or other tissue
sample from the
scene of the crime

2

DNA is extracted from
the cells in the sample
by ultracentrifugation

3

the DNA is cut into
fragments using
restriction enzymes

4

the DNA fragments are
separated by
electrophoresis

5

the fragments are
transferred to a nylon
membrane by Southern
blotting

6

sodium hydroxide is added
to the membrane to split
the DNA into single strands

7 DNA probe

a probe binds to the
DNA on the membrane

8

X-ray film is placed
over the membrane

9

the pattern of the bands
on the X-ray film shows
where the probe has
bound to the DNA

Figure 5.2 Genetic fingerprinting

Not guilty!

The first test case using genetic fingerprinting proved that a suspect who had confessed to a rape and murder was in fact innocent. Two girls had been raped and murdered in the same part of the country, but the crimes were committed three years apart. Investigators suspected a connection between the two and soon a suspect was cross-examined. He admitted to one of the crimes but denied being involved with the other.

Forensic scientists used DNA profiling which showed clearly that both crimes were committed by the same person, but the suspect was not that person. All men from the area were screened and genetic profiles made of 5000 samples. Finally one sample matched those from both crime scenes and the culprit was arrested and convicted.

for analysis. The method simulates the replication of chromosomes, producing millions of copies of DNA in a test tube. It is called *in vitro* **DNA amplification** or **polymerase chain reaction (PCR)**. The technique detects much smaller minisatellites than those involved in earlier attempts, and the pieces of DNA required are almost 100 times smaller than those needed previously.

A problem with DNA is that it deteriorates when left at room temperature for a length of time, making analysis of degraded samples unreliable. Bacterial enzymes break down DNA in the same way as they cause the decay of the rest of a dead cell. Incidentally, this fact makes the *Jurassic Park* idea of recreating extinct dinosaurs just a dream! When using the PCR technique, the DNA fragment analysed is very small so it is unlikely that breaks will have occurred in such a small length. Therefore minute samples can be rapidly and reliably analysed regardless of their age using PCR technology.

Other uses of genetic profiling

The forensic use of DNA profiling is not limited to criminal investigations. Paternity testing has been revolutionised because of the development of DNA analysis, and the technique can also help in the identification of individuals after mass disasters. Many Ukranian and Russian miners were killed when an aeroplane crashed in Scandinavia. Dental records were not available, so DNA analysis was used to match together body parts and then to make a formal identification by comparing DNA profiles with those of close relatives. A similar procedure was carried out to identify victims of the Waco siege in Texas.

A remarkable study resulted in the identification of the remains of Tsar Nicholas II and his family using DNA extracted from their skeletons. They were all reported to have been murdered after the Russian Revolution in 1917 and buried in an unmarked grave. When their suspected grave was eventually discovered, analysis showed that the skeletal remains belonged to a family consisting of a mother, father and three siblings. Prince Philip, Duke of Edinburgh, volunteered to give a blood sample to prove whether the mother shared the same maternal line with him, and the results supported the view that she did. Subsequent extension of the investigation used DNA from the exhumed remains of the Grand Duke of Russia, the brother of Tsar Nicholas II. The results showed that the tsar's remains had been correctly identified. Interestingly, the scientists were also able to show that a woman living in the USA who claimed to be Anastasia, daughter of the tsar, was definitely not!

All the examples of genetic profiling quoted so far involve human DNA. The technique has also been applied to species other than humans. Evolutionary links between groups of similar organisms have been traced by comparing their DNA. Closely related species of small invertebrates are often very difficult to tell apart by external features, and DNA profiling has been an invaluable tool to biologists when identifying new species. The method can also be used to study the extent of inbreeding within populations and consequently make better estimates of the viability of threatened species.

Tracking genetic disorders

The diagnosis of genetic disorders in unborn children has become a very sensitive issue in modern society. As a result of a technique known as gene tracking, together with chorionic villus sampling, it is now possible to find out how potentially dangerous genes are inherited. Chorionic villus sampling takes a small sample of cells from membranes surrounding the fetus in the womb, and enables DNA from fetal cells to be extracted and tested. Armed with this knowledge, prospective parents may choose to continue with an affected pregnancy or terminate it.

In gene tracking, the DNA profile from a particular chromosome of a person who is afflicted with a disorder is compared with the DNA profiles of the parents. If there are sufficient samples of DNA from a family pedigree, partners can be given information relating to the chances of a child being affected by a harmful gene before the child is conceived.

The process used for the diagnosis relies on the same technique already described for DNA fingerprinting. However, the main difference is in the selection of the probe used. Remember that a gene probe is just a portion of single-stranded DNA with a radioactive tracer or label which has been cloned to make millions of copies. Very few DNA probes are complementary to the sequence of bases in a functional gene which actually codes for amino acids. The majority of probes that are used for genetic tracking recognise a sequence of bases very close to the functional gene, but are not functional themselves. These non-functional base sequences tend to be linked to the functional gene and are inherited with them. The closer the marker sequence is to the functional gene, the more useful it is in the diagnosis of a genetic disorder.

Gene therapy

The results of genetic screening pinpoint a potential problem to which **gene therapy** can sometimes offer a solution. In this technique, medical geneticists aim to alter the genes responsible for certain inherited disorders and insert perfect genes into chromosomes to replace the imperfect ones. The new genes enable the affected cells to function properly and therefore remedy the disorder. However, this is easier said than done because of the technical problem of getting the new genes into the specific cells that need them. There are three main methods for introducing the new gene to the chromosomes of recipient cells.

The first method uses viruses to carry genes into cells. One of the first disorders to be treated in this way was adenosine deaminase (ADA) deficiency. ADA is an enzyme vital for making white blood cells in the bone marrow. To treat the deficiency, a sample of bone marrow is taken from the person and the stem cells which normally develop into white blood cells are separated out. Copies of the ADA-producing gene are taken from a human cell and placed inside a virus which is modified so that it cannot replicate. The modified viruses are used to infect the stem cells so that the ADA-producing gene passes into them. Having received new working genes, the stem cells are returned to the patient where they divide and produce a continuous supply of ADA.

Another method of gene therapy is used to treat cystic fibrosis (see page 34). A new gene is introduced to the secretory cells of the lungs by inhaling an aerosol spray containing the new gene. The hope is that the genes will pass into the lung cells and switch on the production of the protein needed for the supply of normal mucus, restoring the functioning of the lungs.

A third approach consists of injecting new genes so that they can reach all the cells of the body. The control units of the genes are programmed to activate the genes only once they are within the target cells. One possible use of this technique is in the treatment of skin cancer (melanoma). Melanoma cells contain a particular section of DNA called a promoter. The promoter regulates the manufacture of an enzyme that controls the production of the brown pigment melanin. Geneticists can now produce genes that are switched on only by the promoter in melanoma cells – in all other cells these genes remain dormant. It might therefore be possible to inject someone with genes that cause cells to self-destruct, in the knowledge that these killer genes would be activated only in the cancer cells.

Food production and genetics

On a commercial scale, humans cultivate approximately 170 different types of food crops and considerably fewer types of domesticated animals for food worldwide. Considering that there are about 250 000 species of flowering plants known to science and about 3500 mammals, we are using only a very small proportion of all the variety available. However, the yield of food per unit area of land has to be high to feed the world's population. It is maintained at a high level by intensive farming, selective breeding and gene technology.

Selective breeding

Since before recorded history humans have been cultivating plants for food by **selective breeding** and the same is true of the domestication of animals. Selective breeding aims to produce plants and animals progressively better suited to human needs. Until relatively recent times this form of breeding by artificial selection was somewhat hit-and-miss. Before the 1860s no one had a clear idea of even the most basic principles of heredity, so the development of an improved strain of plant or breed of animal was by trial and error.

Over thousands of years, hundreds of different varieties of domestic animals and cultivated plants have gradually been developed. Once-wild plants changed almost beyond recognition – cereals, for instance, were developed from wild grasses and cauliflower, broccoli, cabbage and Brussels sprouts all belong to a family which originated as wild plants on the seashore. Similarly all the modern varieties of chickens developed from the jungle fowl of the Far East. The ancestor of the pig is the wild boar, that of the cow the wild ox and that of the over-weight Christmas turkey the wild and slender flying turkey hunted by the original pilgrims of New England, USA.

The discovery of the importance of Mendel's laws of genetics changed the approach of plant and animal breeders, indeed selective breeding today has become the science of applied genetics. The process has become more efficient, but it is still recognised that all the answers are not known and that artificial selection is not an exact science. Thousands of different traits or features may be involved in breeding experiments and therefore the results are rarely as easy to predict as those of Mendel's seven traits in peas.

One of the first of the great pioneers of plant breeding was the American Luther Burbank. He is considered to have been a genius in his work and was responsible for developing a huge list of improved varieties of plants. Perhaps his most famous contribution to agriculture was his success with potato crops. One day in 1871, while examining a field of potatoes in Massachusetts, Burbank noticed fruits growing on one of the plants. This was unusual – although potato plants normally have flowers, they seldom bear fruit. New plants tend to be grown from potato stem tubers rather than from seeds. He saved the seeds from the fruits and planted them. He checked the potatoes growing on the underground stems of the resulting plants and saw that they differed from plant to plant. Some were large and others were small; some produced many potatoes, others produced few. One plant had more potatoes that were also larger and smoother than the others. He selected this plant for future breeding by asexual means. It was named the Burbank variety of potato which soon became popular throughout the USA.

Burbank's success was an example of mass selection, where one plant is chosen for breeding from a large number of individuals. Mass selection is probably the oldest type of artificial selection. The first people to cultivate plants always saved the seeds for the next planting from the plants that produced the best yield. The offspring are then most likely to have the desired traits and, over countless generations, modern crops have been evolved from their wild ancestors.

Mass selection can also produce strains of disease-resistant plants. For example, suppose that a fungus has spread throughout a wheat-producing area. Almost all the wheat has been killed except for two or three plants which survived. Some mutated gene has enabled these plants to resist the fungus. Seeds from these healthy plants are grown the following year. Again the fungus attacks the crop, but this time more plants survive. Over several years this cycle is repeated and each year more plants survive to produce a fungus-resistant strain of wheat.

Closely related strains of the same species can be crossed with the aim of producing offspring showing the best traits of both parents, or **hybrid vigour**. For example, one parent might be chosen because it grows quickly or resists disease. The other parent might have particularly good flowers or fruit. Some of the hybrid offspring will hopefully have all the desired traits. Sometimes the new combination of genes in a hybrid will result in traits not shown in either parent – it may grow larger, have more fruit or resist disease better than either parent.

The opposite situation called **inbreeding** is used to perpetuate desirable qualities. Instead of crossing two parent plants, inbreeding involves the self-pollination of a single parent. In this way, no new genes are introduced from a different plant. Inbreeding is used if a plant breeder is sure that the plant has only the desired genes. The resulting seeds are grown and the offspring are sorted so that only those with the same traits as the parent are chosen to produce seed for the next generation. If the process is repeated many times, the end product is a pure strain which is ready to be used commercially.

An important crop worldwide is maize (corn). A combination of inbreeding and cross-breeding is used in the selection of maize plants and modern hybrids are bred from a double cross involving four different strains of parent plants. Each parent strain is pure bred for some important trait as a result of inbreeding. For example, one strain might be fast growing, another might resist disease, and so on. The hybrid offspring are better than either of the parent strains as many of the desired traits are controlled by dominant genes which tend to mask undesirable recessive genes from the other parent strain.

In order to understand how such a double cross works, we can use letters to represent the parent plants. Let A, B, C and D stand for the four parent plants. During the first cross, A is crossed with B. The male parts are removed from plant B to prevent self-pollination and the female parts of the plant are covered until they are ready to receive the pollen. When they are ripe, they are dusted with pollen from plant A. The seeds that result produce a single-cross hybrid, plant AB. Plants C and D are crossed in the same way to produce a hybrid plant CD.

In the second year, plants AB and CD are crossed in the same way. The seeds produced are hybrid ABCD. These are commercially valuable seeds used to grow crops – the plants grown from them will be ABCD. However, the seeds produced by these plants will not be used as seed for the next generation because new combinations of genes would appear and the second crop could not be relied upon to have the desirable traits of the first.

Engineered plants and animals

Instead of the somewhat hit-and-miss techniques of selective breeding, genetic manipulation can be used to produce ideal crops. The basic techniques for genetically engineering plants are the same as those used in bacteria (see page 129). Fragments of DNA are taken from the donor organism, introduced into carriers and then put into a new host of cultured plant tissue.

There are three main problems encountered with this procedure:

1 It is not always easy to pinpoint the segment of DNA responsible for a particular desirable characteristic.

2 Suitable carrier organisms are not always easy to find. Organisms that can introduce DNA into a plant are often pathogens – they infect the plant and cause disease. A common bacterium used for the purpose is *Agrobacterium tumefaciens* which causes tumours in plants. Such pathogens must be modified so that they will introduce new DNA but will not cause damage.

3 Genes must be controlled when they are in a new host so that they do not spread to wild plants and have any long-lasting unintended effects. This control is usually achieved by modifying the regulatory mechanisms (see page 134).

The potential for the application of genetic engineering to food production is immense. Insects, fungi and other pests compete with us for our crops. The development of DDT and other pesticides led farmers to think that crop yields could be increased by wiping out the pests chemically. They were mistaken. The bugs evolved to become resistant and superimposed on this problem were the toxic effects of the pesticides which became concentrated through food chains, causing alarming potential environmental problems.

Plants with inbuilt pest control systems

What if plants could be persuaded to make their own pesticides? This was a question asked by innovative geneticists in the 1980s and answered soon afterwards. The first commercial crop plant to be genetically engineered to produce its own insecticide was the potato. The insecticide is a toxin produced by the soil bacterium *Bacillus thuringiensis*. It is harmless to humans but kills beetles, caterpillars and most other insect pests including the infamous Colorado beetle. Spiders and many beneficial insects are unharmed by the toxin which also breaks down quickly, so can be considered as environmentally friendly. Scientists isolated the gene in the bacterium that controls the production of the toxin and inserted it into a potato plant so that the plant could make the toxin in its leaves. When pests eat the leaves, they die. Of course, there is a concern that harmful insects may develop resistance to the toxin, and before such a modified plant is used on a large scale it must receive government approval. In the USA the Environmental Protection Agency (EPA) has the final say in such decisions. This particular project was reviewed by the EPA's advisory panel in 1995. The window of opportunity was opened for all other crop plants to be modified in a similar way.

Another ingenious way of combating pests is to make infected plants commit suicide, so that the pest is prevented from spreading. Potatoes have been engineered to behave in such a way if they are attacked by a fungus. Barnase is an enzyme that breaks down RNA and consequently disrupts protein synthesis, killing the plant. Geneticists took a gene that codes for barnase and attached it to one of the potato's controlling promoters. This promoter activates the gene when a pathogen attacks a cell in the plant. The promoter then switches on the barnase gene and kills the cell. Healthy genetically modified plants make small amounts of barnase in their cells, but only when a pathogen attacks a cell will enough barnase be made to kill the cell. The barnase gene is derived from *Bacillus amyloliquefaciens* which is a naturally occurring soil bacterium.

In December 1996 the European Commission approved the sale in Europe of a genetically engineered type of maize which contains genes for herbicide resistance plus a natural insecticide from *Bacillus thuringiensis*.

As part of the engineering process, a marker gene was also inserted which marks the cells that contain the modified chromosomes. The marker gives resistance to a commonly used antibiotic. Critics of the technique suggest that there is a small chance that, when cows eat the maize, the gene could pass into bacteria that normally live in the cow's digestive system. This would help spread antibiotic resistance. Concerns about the presence of the antibiotic-resistance caused the British Government to raise objections to the sale of the modified maize in Europe, but the Commission was forced to consider the financial implications of closing the European market to the product. The USA export hundreds of millions of dollars' worth of maize to Europe each year and to suddenly prevent this could easily start a trade war.

By 1997 governments of those countries which traded in genetically engineered products began tightening up their regulations. The Australia and New Zealand Food Authority (ANZEA) prepared rules which made it much more difficult to sell food made from genetically modified plants or animals. Approval for the sale of such products is subject to meticulous and prolonged scientific scrutiny. Those foods that are approved must carry a label if they contain more than 5% modified material.

Herbicide resistance

As well as insect pests, weeds compete with crops if left unchecked and for a great many years farmers have attempted to eliminate weed growth using herbicides. Selective herbicides that kill weeds but not crops are difficult to find so there is great interest in creating a genetically engineered herbicide-resistant crop. Progress in this work has been rapid, mainly because the growth-inhibiting properties of some herbicides are the same as occur in certain bacteria. A herbicide-controlling gene can be taken from a bacterium and transferred to a plant such as soya bean via a species of *Agrobacterium*. Again there have been protests from people who are worried about the plants having marker genes which confer resistance to antibiotics such as ampicillin. However, tobacco, tomato, potato and rape seed are crops that have already been modified and are therefore obvious choices for commercial production.

Beside the problem of the possible spread of antibiotic resistance, what if herbicide-resistant plants spread throughout the environment? How could they be destroyed if weedkillers cannot work on them? The answer is to ensure that the plants are sterile and can only reproduce by vegetative means.

Up to now we have considered only bacteria as a source of useful genes for giving crops protection. Researchers in California inserted a gene from the snowdrop *Galanthus nivalis* into vines to make them resistant to a harmful roundworm, the dagger nematode *Xiphinema index*. The worm attacks the roots of the vines and also infects them with a virus. The bacterium *Agrobacterium tumefaciens* is used to carry the snowdrop gene into the vine roots, from where they cannot find their way into the grapes.

Improved products

As well as protecting crops from pests and disease through genetic engineering, geneticists turned their attention to using the new biotechnology to improve the quality of plant products, or to make plants produce entirely new products. One example shows how medical technology has made use of engineered plants.

The life-threatening liver disease hepatitis B could soon be treated using genetically engineered bananas modified to carry vaccines. Researchers at Cornell University, New York, have produced bananas which are able to manufacture antigens found in the hepatitis B virus. The carrier bacterium *A. tumefasciens* is used to transfer the gene for the production of the hepatitis B antigen from a virus into the bananas. Theoretically it could provide a very cheap method of vaccinating populations of developing countries and could be extended to prevention of other viral diseases such as measles, yellow fever and polio.

Food technology is an important field today in our world of giant supermarkets, which have to store foods in bulk. This is another area that is benefiting from gene manipulation. The genetically modified Flavr Savr tomato was one of the first fruits to demonstrate the wonders of genetic engineering to the public. Media coverage was widespread when the story broke that scientists had invented a tomato that could last much longer than usual in a fresh and tasty condition. The scientific explanation sounds much less dramatic than the many newspaper headlines and pseudo-scientific reports that appeared in 1994. The tomato has a gene inserted which switches off the synthesis of the enzyme polygalacturonase. It is this enzyme that causes tomatoes to soften when they ripen. The modified tomatoes are less likely to be damaged when they are harvested and can remain longer on the plant to ripen naturally. They should therefore have a much better flavour besides having a longer shelf life.

Geneticists seem to have an unlimited source of innovative new ideas. One of the latest is modifying cotton plants so that they produce fibres containing granules of plastic! Although the idea was tried in 1992 using a type of cress plant, it was not until 1996 that the scope for cotton–plastic fibre production was realised. This new type of fibre can be used to make fabric for ultrawarm clothes, carpets and insulation. Researchers from an American company inserted two genes from the bacterium *Alcaligenes eutrophus* into the cotton plant. This bacterium normally makes a biodegradable plastic called polyhydroxybutyrate (PHB), which is an energy store in the bacterium rather like fat in animals. The genetically engineered cotton plant can make plastic by a modified natural biochemical process that usually produces oils and waxes in the cotton plant. A device called a gene gun was used to fire the two bacterial genes into the cotton plant embryos contained in the seed. First plasmids containing recombinant bacterial DNA were mixed with tiny particles of tungsten. These particles were then stuck on the front of a cylindrical plastic bullet. The device works like a miniature pistol. A firing pin detonates a blank gunpowder charge that propels the bullet down a barrel onto a plate. The impact of the bullet hitting the plate jerks the tungsten and plasmids off the bullet's surface through an opening and across a vacuum to the plant cells. The particles of tungsten are large enough to penetrate cells but they do not destroy the cells.

Nitrogen fixing

Genetic engineering could in future take the place of fertilisers, so avoiding the economic and environmental problems associated with their use. As a rule, plants are adapted to keep microbes out of their tissues. Legumes, plants that produce pods, are exceptions. They welcome invasion by a bacterium called *Rhizobium meliloti* which forms swellings on the roots called root nodules. Both the legume and the bacterium benefit from the relationship – the microbe has a safe home in the roots of the plant, and obtains sugars which have been made by the plant during photosynthesis. In return the microbe provides the valuable service of converting nitrogen from the atmosphere into ammonia, which the plant can use to make proteins. Nitrogen is abundant in the air but is a very unreactive gas – chemists have to subject it to some very harsh conditions of temperature and pressure before they can make ammonia from it. These tiny microbes can make nitrogen react biochemically at normal temperatures. Very few organisms have this nitrogen-fixing ability. Because of this, leguminous crops of peas, beans, clover and alfalfa do not need nitrogen-containing fertiliser – they have their own built-in bacterial fertiliser factories.

Why does *Rhizobium* live in the roots of legumes and not cereal crops? It appears that the bacterium responds to chemicals called flavenoids produced by the roots of legumes. Flavenoids enter the bacteria and once inside trigger the reaction of a key protein which acts as a gene regulator (see page 109). In the presence of flavenoids the protein switches on certain genes in the bacterium. The bacterium makes a chemical signal which passes through the soil to the plant, telling it to make nodules. This relationship between the bacteria and legumes is highly specific – a given variety of bacteria can cause nodule formation only in certain groups of legumes. Legumes in the same group make the same types of flavenoids – for example, peas, vetches and lentils all make one type of flavenoid and so belong to the same group. Clovers belong to a separate group, so clovers cannot normally be invaded by the same strain of bacteria as peas.

By 1997 genetic engineers had succeeded in altering the bacteria that normally invade peas so that they will invade clover. The next stage perhaps will be to enable the bacteria to invade cereal crops. If the genes responsible for the production of the legume flavenoids could be isolated and inserted into cereal crops, this could be one of the greatest breakthroughs in the history of agriculture. Feeding the world without nitrate fertilisers would reduce protein deficiency in populations and also drastically reduce pollution problems resulting from the use of nitrate fertilisers.

A gene too far?

Manipulating genes in plants does not usually promote the same ethical concerns as doing the same with animals. Newspaper headlines like 'Featherless chickens, self-shearing sheep, asexual cattle' have a sensational effect – the idea of monstrosities being purposely made for the sake of scientific research is shocking to the public and genetic engineering will provide material for science fiction writers for many years to come. However, let us look in an objective way at some implications of techniques that are already with us.

Cows, sheep and mosquitoes

In 1996 Rosie, a cow with a difference, was born. Her birth heralded hope for the survival of thousands of premature babies born each year. Scientists genetically engineered Rosie and eight other cows to produce a human protein in their milk. Early in their development they were given human genes which make the protein alpha-lactalbumin, normally found in human milk. This protein is a rich and balanced source of amino acids essential for newborn babies. The protein can be produced in bulk in Rosie's milk, purified and added to powdered milk for premature babies. Typically the breed of cow to which Rosie belongs would produce 10 000 litres of milk per year.

Even before the success with Rosie, researchers had created a sheep with human genes as early as 1993. These animals were modified to produce human proteins in their milk to provide a blood-clotting factor needed by haemophiliacs (see page 115). Alpha-1-antitrypsin is a protein that helps treat cystic fibrosis (see page 34) and this can also be produced in sheep's milk. The 1990s saw other life-saving applications of genetically engineered animals, including an insect – for the first time in 1996 a mosquito that transmits a deadly disease was turned into a harmless if irritating insect by modifying its genes. The disease in this case was encephalitis in children which is relatively rare, but if the same technique could be used against mosquitoes that carry yellow fever and other killer diseases, many thousands of lives could be saved.

Fish that can't find their way home

Most people would see the life-saving benefits of the above three examples of genetically engineered animals. Let us now consider some potential problems with genetically changing animals. One well publicised example is the farmed salmon. It is possible that harmful genes will become concentrated in a population bred in captivity, which could escape back into the wild. Fish farmers have designed a 'super salmon' by artificially selecting genes that control desirable qualities like a high growth rate. In doing this the genes responsible for the salmon's normal homing behaviour have been suppressed. If these salmon escape into the natural wild population and breed, they could introduce mutated genes into the wild population and destroy their normal homing behaviour. The results of a study of this problem have indicated that hybrids between farmed salmon and wild salmon survive less well than native fish during the first four months in the wild. This suggests that native salmon have a degree of genetic adaptation to local conditions which is impossible to simulate in farmed fish.

Giant pigs, featherless chickens, self-shearing sheep

As long ago as 1985, American scientists designed the infamous Beltsville pig that was modified using human growth hormones. It grew so large and so fast that it became crippled with arthritis. As yet growth promotion has not been induced genetically but the driving force of profits in the market-place may fuel the desire to producing fast-growing domesticated animals.

In the 1990s Israeli scientists were working on producing a featherless chicken. The growth of feathers uses up energy in the animal which could be directed towards making more meat, and more meat would theoretically mean more profit. The new technology has hinted at even more bizarre money-making tricks. What if behavioural genes could be altered so that animals become more amenable to intensive factory farming? Such animals would be little more than meat- or egg-producing machines with behaviour changed so that the animals would hardly be recognisable as such.

Australian geneticists have perhaps demonstrated some dangers of this technology. They have produced a sheep with a genetically engineered skin growth hormone that produces breaks in the wool fibres as they grow. All the wool falls off at the same time so that the sheep do not need to be sheared. Unforeseen side-effects were severe sunburn and spontaneous abortions.

Organ transplants

For many years there have been attempts at transplanting non-human organs into humans. The poor success rate has been due to rejection of unmatched tissues, which is a problem even between humans. In an effort to overcome this problem pigs have been developed that contain human genes. Early in the next millenium 'spare part' pigs could be available with kidneys, hearts and lungs ready to be donated to human recipients.

Cloning animals

Selective breeding has long been used on farm animals. Since the 1950s artificial insemination techniques have allowed a single bull to father thousands of calves, but their genetic make-up still varied quite widely because of differences in the mother cows. The present quest for genetic perfection has developed a new science of cloning. The result could be a super-race of genetically identical animals, tailor-made to produce any protein we want on demand or to produce meat of optimal quality. However, this super-race would all be identically vulnerable to the same diseases which could eliminate entire herds, as has happened with cloned super-crops.

Mass-production by cloning is well on the way to reality in many countries. In 1997 Australian researchers created almost 500 genetically identical cow embryos. The research leading to the creation of Dolly the sheep (see page 126) may lead to hundreds of copies of adult animals being made from a single cell. For example, an elite cow's egg could be fertilised with a prize bull's sperm and then hundreds of genetically elite offspring produced by cloning and implanting the eggs in surrogate cows.

Will the next step be the cloning of humanised pigs or sheep to supply human blood? This would eliminate the need for human blood donors and provide a bottomless blood bank. Or perhaps we have already gone a gene too far!

APPENDIX
A SHORT HISTORY OF GENETICS

384 BC Aristotle's *Politics* advocated the first selective breeding of humans and therefore the first attempt at eugenics.

1630 Huntington's chorea introduced to North America by immigrants from Suffolk, England.

1665 Robert Hooke published his *Micrographia* with the first descriptions of cells.

1669 Johann Joachim Becher observed and recorded the results of purposely mixing characters breeding pigeons and also breeding trees.

1700 Antoni van Leeuwenhoek first described the nucleus of a cell.

1780 Lazzaro Spallanzani demonstrated that sperms are needed to fertilise eggs.

1819 M H Braconnet was the first to isolate amino acids.

1822 John Goss recorded results of crossing green-seeded peas with yellow-seeded peas in the *Journal of the London Horticultural Society*.

1835 Jöns Jakon Berzelius was the first to conceive the idea of an enzyme.

1838 Gerardus Johannes Mulder first coined the word 'protein'.

1839 Theodor Schwann and Matthias Jakob Schleiden confirmed that the cell is the basic unit of life.

1850 Chromosomes were first described by Friedrich Wilhelm Hofmeister.

1865 Gregor Mendel presented his research paper *Experiments in Plant Hybridisation* to the Brunn Natural History Society.

1866 Down's syndrome was first described by Langdon Down.

1869 Friedrich Miescher discovered nucleic acids; he called them nuclein.

1871 Luther Burbank pioneered the selective breeding of plants and produced the Burbank potato.

1872 George Huntington first described the disorder Huntington's chorea.

1875 Oscar Hertwig first observed fertilisation.

1881 Edouard Balbiani first described giant chromosomes in the salivary glands of *Drosophila.*

1882 Walther Flemming first described mitosis.

1883 Robert Brown established the idea that a nucleus is an essential part of cells.

1883 Francis Galton was the first to coin the term 'eugenics'.

1887 August Weismann first described meiosis.

1900 Hugo de Vries, Carl Correns and Erich Tschermak rediscovered the work of Mendel and realised its significance.

1900 Karl Landsteiner invented the ABO system of blood grouping, which was a prerequisite for understanding the genetics of blood groups.

1901 Emil Fischer worked out the chemistry of purines.

1903 Walter S Sutton suggested that chromosomes carry genetic information.

1905 William Bateson and Reginald Crundall Punnett discovered linkage of genes.

1905 Punnet invented the Punnet square method to predict the outcomes of genetic crosses.

1905 Lucien Cuenot first discovered lethal genes in mice.

1907 William Bateson first coined the word 'genetics'.

1907 Eduard Buchner received the Nobel prize for chemistry for demonstrating that enzymes from yeast could work as extracts.

1908 Sir Archibald Garrod suggested that genes make enzymes.

1909 Wilhelm Johannsen first coined the word 'gene'.

1909 The crossing over of chromosomes during meiosis was first described by Frans-Alfons Janssens.

1910 Albrecht Kossel received the Nobel prize for medicine for identifying part of the nucleic acid molecule.

1910 Aaron Theodor Levene identified ribose and deoxyribose sugars in nucleic acids.

1915 The first primitive genetic maps were made by Alfred Sturtevant.

1916 Calvin B Bridges proved that genes were associated with chromosomes.

1924 Robert Feulgen established that DNA is confined to chromosomes.

1928 Frederick Griffith demonstrated that genetic material could be passed from one bacterium to another.

1930 D Kostoff rediscovered the work of Balbiani (1881) who first described giant chromosomes. Kostoff used the information for genetic mapping.

1931 The USA introduced sterilisation laws aimed at improving the population by eugenics.

1933 Thomas Hunt Morgan received the Nobel prize for medicine and physiology for discovering sex-linked genes in *Drosophila*.

1933 Theophilus Schickel Painter was the first to make accurate physical chromosome maps.

1934 Genes were measured for the first time by Hermann Joseph Muller and A A Prokofyeva.

1937 George Wells Beadle and Edward Lawrie Tatum demonstrated that genes have a chemical effect on the development of organisms.

1938 Torbjörn Caspersson demonstrated the link between nucleic acid and the synthesis of materials in cells.

1939 Karl Landsteiner discovered the Rhesus factor in blood. This was a prerequisite for the understanding of the genetics of blood groups.

1940 Adolf Hitler attempted to produce the 'master race' by eugenics.

1944 Oswald Theodore Avery, Colin Munro Macleod and Maclyn McCarty showed that DNA is responsible for passing information from one generation to the next.

1947 Erwin Chargaff analysed DNA and determined the relative amounts of its four bases.

1947 James Batchellor Sumner shared a Nobel prize for chemistry for being the first to crystallise and isolate a pure enzyme.

1949 Murray Barr first discovered Barr bodies which are used in gender determination.

1949 James Van Gundia Neel was the first to work out the inheritance of sickle-cell anaemia.

1950 Linus Pauling and Robert Brainard Corey discovered the helical nature of some protein molecules.

1952 Alexander Lathan Dounce predicted a three-base codon in DNA.

1953 Francis Harry Compton Crick and James Dewey Watson discovered the structure of DNA.

1956 Tjio and Levan determined that there are forty-six chromosomes in human cells.

1957 Alexander Robertus Todd received the Nobel prize for chemistry for synthesising chemicals leading to the discovery of the structure of DNA.

1958 Matthew Stanley Meselson and Franklin William Stahl developed the semiconservative theory of replication of DNA.

1958 Beadle and Tatum received the Nobel prize for medicine and physiology for demonstrating that one gene controls the production of one enzyme.

1959 Arthur Kornberg received the Nobel prize for medicine and physiology for demonstrating that DNA can copy itself.

1959 Klinefelter's syndrome was described as XXY. The trisomy of Down's syndrome was also discovered.

1962 Crick and Watson received the Nobel prize for medicine and physiology for their discovery of the structure of DNA.

1963 François Jacob and Jacques Monod described the role of messenger RNA in protein synthesis.

1965 Jacob and Monod received the Nobel prize for medicine and physiology for demonstrating how genes are switched on and off.

1966 The genetic code is understood.

1975 At Asilomar, California, rules were formulated to eliminate the possibility of the escape of genetically engineered organisms into the environment.

1977 The Dounreay nuclear power station explosion occurred and research was done into the possible effects of radiation on genes.

1985 The first extensive human gene maps were made.

1986 The Chernobyl nuclear power station disaster took place and the potential for massive increases in the frequency of gene mutation in local populations was discovered.

1988 Genetic fingerprinting was first used in Britain to convict a murderer.

1990 Lap Chee Tsui discovered the defective gene for cystic fibrosis.

1993 The exact position of the gene for Huntington's chorea was discovered.

1993 Transgenic sheep were developed to produce human proteins in their milk.

1994 Genetically engineered tomatoes that keep fresh longer were developed.

1995 The Criminal Justice Act legalised a national DNA database to store genetic profiles of all convicted criminals.

1995 China introduced a eugenics law to make screening mandatory for 'inappropriate' genes.

1996 Cloning of sheep from different cells of the same embryo was successful.

1996 The European Commission approved the sale of maize that was genetically engineered to make its own herbicide.

1996 Cotton was genetically engineered to produce plastic and which gave the fibre better insulating properties.

1996 Rosie the genetically engineered cow produced human protein in her milk.

1997 Australian researchers cloned 470 genetically identical cow embryos from a single embryo.

1997 The first recommendation for a nationwide genetic test was made by the National Institutes of Health in the USA.

1997 Dolly the sheep was produced from the cloning of an adult cell rather than from cells of an embryo.

GLOSSARY

Allele   One of two or more alternative forms of a gene, bringing about contrasted genetic characters.

Amino acid   A building block of a protein molecule.

Anticodon   The triplet of nucleotides in a transfer RNA molecule pairs with a specific triplet codon in messenger RNA during protein synthesis.

Asexual reproduction   Reproduction that does not involve fusion of gametes.

Autosome   Any chromosome other than the sex chromosomes (X and Y).

Bacteriophage   A virus which infects a bacterial cell and reproduces inside it.

Barr bodies   Dark staining features in the nuclei of the cells of female mammals, representing the condensed X chromosome.

Catalyst   A substance that increases the rate of a reaction without being used up in the reaction.

Cell membrane   The outer boundary of the cytoplasm of a cell.

Cell theory   The theory that all organisms are composed of cells and cell products and that growth and development results from the division and differentiation of cells. It is generally attributed to Schleiden and Schwann in 1839.

Cell wall   A non-living layer of cellulose surrounding the cells of plants and fungi.

Centrioles   Small bodies present in pairs in some cells, which are important in forming the spindle during cell division.

Centromere   A specialised part of a chromosome which attaches the chromatids together during the early stages of cell division.

Character   Any observable phenotypic feature of a developing or fully developed individual.

Chiasma   The cross-shaped arrangement of the chromatids which is formed at their point of exchange during crossing over in meiosis.

Chromatid   The two parts of a chromosome formed when the chromosome is seen to be split in half lengthways at cell division.

Chromosome   One of the threadlike structures present in the nuclei of cells, carrying genes along its length.

Clone   A collection of genetically identical organisms or cells descended from a single ancestor by asexual reproduction.

Cloning (molecular)   The production of a number of identical DNA molecules by replication of a single DNA fragment in a suitable host system, such as a bacterial plasmid.

Codominance   Two or more alleles making a positive contribution to a phenotype resulting in blending of characters.

Codon   A triplet of nucleotide bases in DNA that codes for a specific amino acid in protein synthesis.

Continuous variation   A type of variation which shows an even graduation between two extremes in a population.

Contrasted characters   Pairs of characters, one of which is dominant, the other recessive.

Crossing over   A process of exchange between paired chromosomes which may give rise to new combinations of characters. It takes place by the breaking and rejoining of chromatids and leads to the formation of chiasmata.

Cytoplasm   The substance of a living cell excluding the nucleus.

Dihybrid   A hybrid heterozygous at two loci and obtained by crossing homozygous parents with different alleles at two given loci: for example, Mendel's cross between yellow round (YYRR) and green wrinkled (yyrr) garden peas to give a yellow round dihybrid (YyRr).

Dihybrid inheritance   The inheritance of two pairs of contrasted characters.

Diploid   A cell or organism with two sets of chromosomes, each chromosome having a partner.

Discontinuous variation   A type of variation in populations where the individuals fall within two or more distinct groups with respect to a particular character. This type of variation usually occurs where there is a completely dominant and a completely recessive character in a pair of alleles.

DNA   Deoxyribonucleic acid – one of the components of chromosomes and the chemical which makes up genes. It is a double-stranded molecule made up of units called nucleotides.

Dominance   A character is said to be dominant if the allele controlling it produces the same effect in the heterozygous as in the homozygous state.

Double helix   Two spirals wound around each other in opposite directions.

Environment   The complete range of external conditions within which an organism lives, including physical, chemical and biological factors such as temperature, light and availability of water.

Enzyme   A protein which alters the rate of chemical reactions.

Equator   The plane through the middle of a cell, away from which the chromosomes move at cell division.

Eugenics   The theory that the human race could be improved by controlled selective breeding between individuals with 'desirable' characteristics.

Feedback control   The regulation of the formation of a product of a chemical reaction by the accumulation of the product above an optimum amount. However, once the product is utilised, there is no inhibition and the reaction proceeds until the product reaches an optimum once more.

Fertilisation   Fusion of gametes in sexual reproduction.

Filial generation   Offspring of a mating.

Gamete   The cells (egg and sperm) which fuse to form a zygote in sexual reproduction.

Gene   The genetic material responsible for determining the structure of a particular protein or chain of amino acids; it represents Mendel's 'germinal unit'. Genes determine all inherited characters.

Genetic code   The means by which information is carried in the genetic material to control protein synthesis.

Genetic engineering   The production of recombinant DNA molecules by the insertion of donor DNA into a virus or bacterial plasmid which is incorporated into a host organism.

Genetic material   Something handed on from generation to generation which partially determines an organism's phenotype.

Genetics   The science of heredity – the study of the transmission, structure and action of the material in the cell which is responsible for inheritance.

Genome   The total genes in a basic set of chromosomes of an organism.

Genotype   The genetic make-up of an organism with regard to a given pair of alleles, for example a tall pea plant's genotype could be **TT** or **Tt**.

Haploid   The possession of only one set of chromosomes. The gametes are haploid.

Heterozygote   An individual who received unlike alleles from its two parents for a particular characteristic.

Homozygote   An individual who received similar alleles from its two parents for a particular characteristic.

Hybrid   Term used by Mendel to mean any offspring produced by sexual reproduction between two visibly different parents; usually taken to mean an offspring of two parents of different varieties of a species.

Independent assortment (principle of)   The principle, formulated by Mendel, that genes segregate independetly at meiosis so that any one combination of alleles is as likely to appear in the offspring as any other combination.

Karyotype   The chromosome complement of an individual defined by the number, the form and the size of its chromosomes.

Ligase   An enzyme which closes breaks in single-stranded

DNA.

Linkage   The association of certain genes in their inheritance.

Linkage group   A group of gene positions in the same chromosome.

Linkage map   A chromosome map showing in linear order the genes belonging to the linkage group for that chromosome.

Locus   The site in the chromosomes where a gene is located.

Meiosis   Cell division in which the chromosome number is halved so that the diploid cell gives rise to haploid cells (often gametes).

Messenger RNA (mRNA)   A single-stranded RNA molecule formed by transcription from DNA which carries the information encoded in the gene to the site of protein synthesis on the ribosome.

Mitosis   Cell division in which daughter cells have the same number of chromosomes as the parent cell.

Monohybrid inheritance   The inheritance of one of a pair of contrasted characters.

Mutation   A change in a gene or a chromosome, which may result in the conversion of one allele to an alternative allele or in rearrangement of the genes on a chromosome or in an abnormal number of chromosomes.

Natural selection   The selection of survivors by selective pressures of the environment. Those with variations best suited to environmental change survive to breed and pass on the variations to their offspring.

Non-disjunction   A failure of chromosomes to separate during cell division.

Nucleolus   A structure in the nucleus associated with coding for ribosomal RNA.

Nucleotide   The basic unit of nucleic acid molecules composed of a phosphate, a sugar and a base.

Nucleus   A membrane-bound part of a cell that contains chromosomes.

Operon   A group of adjacent structural genes along with operator and promotor sites which switch the genes off and on.

Phenotype The outward expression of a gene – the visible characteristics shown by an organism.

Plasmid A circular DNA molecule in a bacterium that replicates independently of the main chromosome.

Pollination The transfer of pollen from the anther to a stigma of a flowering plant.

Polymerase An enzyme which catalyses the assembly of nucleotides into RNA or DNA on a DNA template during transcription.

Polypeptide A compound that contains many amino acids linked together with peptide links.

Protein An organic compound composed of chains of amino acids linked together by peptide bonds.

Punnet square A method used to determine the probabilities of allele combinations in a zygote.

Pure strain Homozygotes that are identical to each other and which continue to breed to produce genetically identical offspring.

Recessive A character is said to be recessive if the allele controlling it only produces an effect in the homozygous state. The allele does not show itself in the heterozygote but can be passed on unaltered to future generations.

Recombinant DNA A molecule of DNA made from two pieces of different DNA joined together.

Recombination The regrouping of genes as a result of crossing over during meiosis.

Replication The synthesis of DNA.

Repressor A protein that binds to an operator site and prevents transcription of an operon – it prevents the gene being expressed.

Restriction enzyme An enzyme that makes cuts in double-stranded DNA at specific sites.

RNA Ribonucleic acid – a single-stranded molecule made of nucleotides. It exists in three forms: messenger RNA, transfer

RNA and ribosomal RNA.

Ribosome   A structure in a cell composed of RNA and proteins, where messenger RNA is translated and proteins synthesised.

Segregation   Separation of the alleles in the gametes.

Semiconservative replication   A theory to explain the replication of DNA where two identical molecules result by the original DNA molecule unwinding and synthesising two identical daughter molecules. The newly formed daughter molecules each contained (or conserved) one of the original parental strands, plus one newly synthesised strand.

Sex chromosomes   Chromosomes represented by the letters X and Y which determine the gender of an individual. In humans a female is XX and a male is XY.

Sex linkage   A gene is said to be sex linked when it is on either the X or Y chromosome. If it is on the non-pairing part of the Y chromosome it can never cross over to the X chromosome and will therefore always be passed from father to son. If a sex-linked gene is on the X chromosome, a man will pass it on to his daughters and a woman will pass it on to either sons or daughters.

Species   A group of organisms, the members of which interbreed to produce fertile offspring.

Spindle   A fibrous structure which appears at division in some cells and controls the movements of the chromosomes.

Template   A molecular mould for the synthesis of a complementary molecule.

Test cross   A cross to determine the genotypes of identical phenotypes. For example, to distinguish between two identical phenotypes AA and Aa, they can be both back crossed to a known genotype aa and the offspring ratios can be predicted. AA x aa will always give offspring showing the dominant character, whereas AA x aa will give some offspring showing the recessive character.

Transcription   The process in which messenger RNA is synthesised from a DNA template.

Translation   The process by which the information carried in the base sequence of messenger RNA is used to produce a sequence of amino acids in protein synthesis.

Variety   A group of individuals within a species having one or more distinctive characteristics, but generally able to breed freely with all members of the species.

Zygote   A fertilised egg.

INDEX